Henry Alford, Elizabeth Mary Alford

Fireside Homilies

Henry Alford, Elizabeth Mary Alford

Fireside Homilies

ISBN/EAN: 9783337253578

Printed in Europe, USA, Canada, Australia, Japan

Cover: Foto ©berggeist007 / pixelio.de

More available books at **www.hansebooks.com**

FIRESIDE HOMILIES

BY THE LATE

HENRY ALFORD, D.D.

DEAN OF CANTERBURY

EDITED BY HIS WIDOW

NEW YORK:

ANSON D. F. RANDOLPH & COMPANY,

770, BROADWAY.

COR. 9TH ST.

PREFACE.

A T the request of the Editor, I have undertaken to write a preface to the following
Homilies, explaining her motives for repub-
lishing them, and the circumstances under
which they were written. They appeared first
in the *Sunday Magazine* during the years
1868—1871; the last two in the number for
January, 1871, the very month in which, after
so short an illness, the busy life of their author
came to its peaceful close.

It had been his intention to republish them
in a separate form, and with this view a
portion of them was already in the printer's
hands at the time of his death. His widow,

therefore, looks upon the carrying out of this intention of his as a sacred trust.

Moreover, these Homilies are, as a friend has remarked, " more like autobiography than anything he has left in prose—a sort of autotype of what he was in family life at the fireside ;" and it is in this light also, as a fitting supplement to his " Life," that the Editor has been anxious for their republication.

It is needless here to enter at any length into the character or acquirements of Dean Alford, the former having been already set forth in all its purity, single-mindedness, and earnestness, and the latter in their versatility and extent, in the " Life " already referred to. But there is one aspect in which it may be well for a while to regard him before entering upon the following Homilies, and that is, by the light of his own fireside, in the atmosphere of his home.

Few men, we should think, ever had the feeling of family affection more strongly developed in them than he had. So intense was it, indeed, that at times it seemed almost to border upon pain. Thus the natural breaking-up of the family circle was a real trial to him, and it is to the void felt at his fireside after the marriage of his second daughter, that we owe these touching Homilies. We who remember the happy Canterbury Sundays of the past, before the family party was broken up, can easily picture those later ones on which the following pages were written. The daughters, who were so much to their father, and the cousins and friends they gathered around them, were no longer grouped about in the large drawing-room, he, the centre of them all. The many-sided chat that comes from the contact of many minds was no longer possible. He and his life's companion were alone once more —as much alone as when she left her father's home at Heale, and they went forth to face the world together, in the spring-tide of their youth. But there were no gaps in the family circle then, no missing ones to mourn. Now, after the lapse of years, the case was altered. Two graves in Wymeswold churchyard had long ago claimed two missing ones, two happy hearths elsewhere had made two more inroads on his own. And so he was fain to call around his chair on these later Sunday evenings the "brain children" of his fancy, and pour out to them some of the bright, fresh thoughts, that floated through his ever-active mind in these "Fireside Homilies."

To those who knew him well they are eloquent of many lovable traits in his character. Here we see his hungering after sympathy, in this calling of his dear ones around him to be partakers of his choicest thoughts. Here we see his fearless truthfulness, in neither slurring over anything that was, nor pretending to anything that was not his own sincere conviction. Here, too, we catch a glimpse of that wonderful buoyancy of his nature, which threw such a charm over his life. This perhaps was never more evident than a few weeks before his death, when, suddenly debarred by the doctor's decree from literary work—that work which had seemed an essential part of his existence—not even a shadow appeared to fall on his spirit; but, as though braced-up by the sudden trial to a still higher stand-point, we find him full of thankfulness for the work-time already allowed him, full of bright resources for the future.*

But it was not only in this rebound after a blow, which would have been crushing to many men, that the buoyant youthfulness of his nature showed itself. With children he was a child again, romping with them so much on their own level, that reverence was apt to be forgotten in the fun. To young people he was especially kind, throwing himself so heartily into their pleasures and pursuits, whether it were a picnic, a charade, a game of croquet, or a country walk, that they felt

* See the " Life," 3rd edition, p. 469.

him to be one of themselves. And at the
same time he was ever ready to draw out
and encourage any taste or talent that might
be latent in them. Most justly does Dr.
Stoughton remark: "Dr. Alford's nature was
formed for kindly and loving companionship," *
and most cheerfully does this come home to
us in these " Fireside Homilies."

Here we see too, in dainty bits of word-
painting, traces of his great love of nature.
The same friend, Dr. Stoughton, touches on
this point also, in his description of a walk he
took with the Dean at Canterbury in January,
1869.† And even as we write we recall,
with something like a present enjoyment of
them, delightful rambles in that pretty neighbourhood in bygone days. The time for these
long walks was usually between the close of
the afternoon service in the Cathedral, and the
seven o'clock dinner-hour at the Deanery.

* See " Life," p. 419.
† Ibid., 3rd edition, p. 512.

Then, a large family party, we would go off to the surrounding woods, the Dean entering so heartily into the pleasure of it all; ever ready to point out any fine view or striking feature in the landscape, sometimes sitting sketching while we roamed about, always taking an interest in the botany of the woods, and often returning home laden with spoils.

But perhaps his enjoyment of art comes out still more strongly in the following Homilies. How lovingly he dwells on each detail in his favourite pictures! We are told that it needs a poet thoroughly to appreciate poetry, and so surely it is with painting. In Dean Alford's case this need was fully met. He was an artist himself of no mean capacity, and traces with an artist's eye the most striking merits in these works of the old masters.

Something more, however, was necessary for their full interpretation. The deeply religious sentiments embodied in these works required above all a deeply religious mind to enter into them, and this was a qualification pre-eminently possessed by the Dean, as his "Life," from first to last, testifies. It is, indeed, for the insight we gain, through his "Journals and Letters," into the earnestness of his inner Christian life, from his boyhood to his death, that we especially prize them. Trusting that the same earnest, loving spirit, may make itself felt in these "Fireside Homilies," we commend them to the public, reminding them, in his own words, that these "evening talks" of his are "rather exercises of the fancy about divine things, than regular treatments of divine things themselves." *

E. M. ALFORD.

The Mount, Taunton,
September, 1874.

---

* See "Fireside Homilies," No. X. p. 151.

I.

HERE we are, darlings, by our cozy fireside this Sunday evening. If there be a heaven on earth, it is this. Very little makes it up. We have here no scenery, no gaiety, no merry game going on: yet how happy we all are! There is dear mamma's loving face, one half in the bright glow of the fire; and Jessie's curls glittering in the same; and little frail Margey, holding up her book to screen her cheek, and her other hand resting with mine on my knee; and there are all the loved ones far away but seeming very near to us to-night. And why are we so happy?

Well, it's because we *believe*. We, sitting here on the day of rest, and at the end of its services, are just enjoying the *luxury of faith*.

B

What an odd word, Jessie, is not it ? But it's a very true one.   I said, and I say, the *luxury* of faith.   And how so ?

Why first, perhaps, because the day's *labours* of faith are over.   I have done my work, and you have done yours.* My two services, and afternoon school, and christenings, are over ; and just as we feel, after a good walk, a glow that keeps the heat in us for hours, so I feel my heart, and my body too, aglow with the day's exercise.   So much good talking, so much honest ministering, have warmed up my religion, which on week days is often so cold ; and I'm like an old dog that has been hunting all day, curled round before the master's fire. And you too, darlings, you've done your work. You've some of you, no more classes to teach ; you've all of you, no more sermons to listen to.   I dare say you know the feeling of waking on a Sunday morning with a kind of dread—

* His thoughts are here probably returning to the days of his parish work at Wymeswold, Leicestershire.—ED.

a feeling of something unusual about to be that day; and then you suddenly remember that it's Sunday, and that's some relief, but still somehow you wish it were a week-day. But then how much better it turns out than you had apprehended! Even in the taking out of the clean things to put on, even in the handling of the last new dress, there is a trifling pleasure which reconciles us to the day; and then as it goes on, the further in, the pleasanter. You sometimes think you'd like to sit at home instead of going to church; those dull prayers—that long dreary sermon —how I wish it would pour, and we couldn't go! But then when you get there, you gradually get interested; some saying of our dearest Lord gets down into the well of the heart, and brings up the fresh spring water to the eyes; or the singing cheers you; or the sermon touches you; and you come away with a feeling which you don't express, that it was good to be there.

And yet, with all this, who doesn't feel glad that Sunday's services are over? They were very good for us; they lifted our thoughts, and inspired our nature; but lifting and inspiring are tiring processes for the same poor nature, and so at the end of them there is a kind of cheery satisfaction that the day is done.

But don't think that this satisfaction is an unchristian or an unbelieving one. I think it is just the contrary. We feel quiet and satisfied after a good meal. I, for one, like to lie back in this old chair and feel one thumb against the other. But it isn't from any disrespect to beef and pudding—far from it; it's because I'm digesting them, and am thankful for them, and satisfied with partaking of them.

And so it is now. The holy faith, to us four who sit round this fire, is in us, digesting. It has calmed and fed us. All its mastication, so to speak, is over for the time. Its questions, and doctrines, and exhortations, and warnings, which we have been working at all day, are at present out of our sight. And we are just sitting in the light of our Father's countenance, enjoying his love to us, as we shall sit when this great religious service of Life is over, and we have taken our places round the fireside at his Home.

Well, this is one reason. But the effect which the day has had upon us may be looked at in other lights. What we have done and shared in to-day, has had an *actual* effect on our religious being. It would hardly be possible for so many hours to be talking about and making real the objects of our faith, without that faith having become more sure, and more real, than it was before. Don't you, my darlings,—I am sure I do,—find it inexpressibly difficult on a week-day to think of the great throne, and the sea of glass, and the blessed elders, and the glorified Lamb, right over our heads, and of ourselves as in their

presence ? Suppose I am riding, or swimming,
or writing letters, or doing any common thing,
such a thought pulls me up as it were with a
jerk ; and I'm obliged to stop a moment and
think ; and before I can really imagine them
above me, a page of Scripture seems to rise
before my eyes, and I see them through *it*.
Not so this evening. Here they all are—around
us—above us. We have been, like the Apostle,
on the house-top praying, till we have fallen
into a trance. Faith has had its swing for
a whole day, and we are penetrated with it.
I have often thought I should like to die on
a Sunday evening, after the services of the
day ; the change would be so small. The
world sounds so distant to-night. It isn't in
the streets ; it isn't on our tables ; it isn't on
our lips. There is no look of it on our faces.
Our rooms, and our furniture, and our fire-
place, and its bright living inhabitant, all
seem only as the cabin of the ship in which
we are taking our voyage. Now at other times

I find myself hammering away at this cabin, painting it, and gilding it, as if I were to be in it for ever; but on these evenings, I am lifted out of it altogether; I can sit in it, and think on the haven where I would be.

And what a thing it is thus to get into tune—thus to have at last succeeded in fixing the focus of faith's telescope, and to see the bright hills, and the glorious trees, and the sweet streams, and the clustering pinnacles, of the land far away! O what a Father we have! What a price He paid for us, and how He loves us! Here are we four—nothing on earth can ever harm us—dear little Margey may have as much sickness to go through as she has gone through already—dear Jessie may be disappointed in what we know is so near her heart—mamma and I may have many a rough place in our sloping path before we come to the river beneath; but here we feel, this Sunday evening, that all will be as nothing: the mighty Love, the everlasting arms—these will keep us safe; our great Sunday evening will be as peaceful, and as cozy, as this, one of our little Sunday evenings, ay, and a thousand times happier too.

But then, darlings, we ought not to think of ourselves only. Even little Margey knows, that to look at a very beautiful thing is intense delight. And what is so beautiful as that which we have been looking at all day—which we can look at quietly, and as if it were our own, to-night? I mean the whole great gift that our Father has given us—ourselves; and the great world to live in and die in, and the sweet life that is to come, and the blessed, loving, and perfect One, in us and about us, and waiting there for us? What a city of beauty, what a school of beauty, what a great cathedral of beauty it all is! Every turn, every corner, every shifting light, have new objects full of light and glory. What hundreds of poets have sung of it—what hundreds of painters have got their subjects from it—what hundreds of musicians have tuned their lyres to it! What art is like Christian art—has such noble forms, such glorious scenes of suffering and of acting, such divine faces—such blessed repose in the scenes of earth—such flights of imagination in the anticipating heaven? What nature is like Christian nature—soft, without effeminacy—pure, without prudery—free, and yet obedient? The very point of union of reclaimed nature and glorified art is the body of the Incarnate Redeemer. In its purity of sinless, unclothed childhood—in its majesty of teaching and healing—in its mighty power of suffering—in its victory of resurrection glory—Jesus, the Son of God, the image of the Father, the spotless, the divine, has brought into our world the form of perfect beauty, for all to look upon and be blessed.

Then again, to come down from the Highest among ourselves again, there's another reason yet why we are so happy this evening. I dare say, little Margey, you would like often to come and nestle close to me as now; but then you look at papa, and you see a great book full of figures before him ; or you see him with his envelopes and his note paper, looking puzzled, and you know he's *busy* (terrible word for you, poor little lassie); and if you were to come and lay your hand in his, you're afraid he would look perhaps sterner than usual and say, " Don't, my child, I'm *busy*." But to-night you've no such fear.   You know the little hand will be clasped, and put on the knee ; there's no such word as " *busy* " to come between papa and you.   And if mamma, or if Jessie ventures on other days to sit cozy by the fire, the thought immediately comes, " Ah, but I ought to be working this or writing that; this is very comfortable, but really I must be up and doing, lazy soul that I am ! "

Now on these evenings there are no such thoughts to interfere with *rest*.   Rest is our right, and we take our fill of it.   It is an approach, my darlings, to the *state of the blessed dead,** when they have earned their repose and sleep, "full of rest from head to feet," as the poet has it. The pleasure of other days becomes the duty of this one ; and so there is no contrast between duty and pleasure, and duty becomes sweetness itself.

Well, my children, see there's dear mamma dropped to sleep, and carrying the *rest* a step further than we are ; and *I* have been actually preaching you another sermon, when I said we were happy because there were no more sermons to hear. But I know it's been *from* the heart: and I think by Jessie's bright swimming eye, and by Margey's pressing hand every now and then, that it was spoken *to* the heart.

And now, Jessie, my love, for that glorious

---

* The Dean at this time, Advent 1868, was preaching in Canterbury Cathedral a series of sermons on "The State of the Blessed Dead," afterwards published in a small volume. —Ed.

"Comfort ye my people." I'm afraid my tenor won't reach that G, after all these sermons. Never mind, there'll be the magnificent chorus, " All flesh shall see it together."

Oh, when, my darlings, when?

II

SO I must do it again, Margey, must I?
Well, only mind one thing, my little one.
If a thing's worth doing, let us do it again
and again ; but don't let us fall into habits for
habit's sake. I've a horror of things becoming
institutions, merely because we love to do the
same thing at the same time. And if I find
myself getting into this (mind, it's a temptation we are all liable to), I make a point of
altering the time, or the manner, or something, so that the act may not be the same,
or may not be done at the same time. You
know, some things must be thus done ; for
instance, meals, and prayers, and recurring
celebrations of any sort : but our tendency is,
to weave nets round ourselves which Provi-

dence has not woven; and then if anything breaks these nets, we make ourselves miserable about it. What do you think about it, Jessie? for I know mamma is with me.

"Well, papa, bondage is very pleasant; far pleasanter than freedom. You've got what to do set you already, and haven't the trouble of choosing; and then it seems so important, and such a duty when it comes. And when you go to bed, you think yourself so good for having been punctual all the day, when perhaps there was nothing worth being punctual about."

All right, Jessie; that rings well. But we mustn't lose our homily. And now what shall it be about? I know, I think—and little Margey is squeezing my fingers so hard, that I think she knows too.

What a subject it is! All Saints' Day! Well, suppose we don't care about commemorations, as some good Christian men don't. And suppose some one said to us, "Let us talk now about the great multitude whom no man can number——about those who are before the throne of God, and have all tears wiped from their eyes,"——what eye of ours would not sparkle, what spirit would not glow with delight? For surely, darlings, it is the subject of all subjects to talk of and to look into. Here are we four, loving the dear Lord, and trusting in Him——and we seem all alone in a corner of the great world; we see very few besides one another; we live mostly in private, and a happy life it is. But how strange it is to think, that the happiest life of all——far more blessed than anything we know of here below——will be a life in public! What an intolerable life it would be to us now! Only fancy dear mamma turned out of her snug boudoir, with her books and her drawing, and her working for the poor, into a great multitude whom no man can number! There was a day when Jessie perhaps would have liked it, but I don't think she would now that one eye has singled her out and loves to look into hers.

And then think of this, too—that this supreme happiness of being among the innumerable multitude will be enjoyed by every one in it. How immensely men must be altered, before that can be the case! Surely there must be some dark poison mixed among the common blood of humanity, which makes men hate one another. In the present life, this hate is almost a necessity, at least in its passive forms. It guards us and our plans and our feelings from being tossed about upon men's tongues and made sport of. It prevents the knowing and interfering from taking advantage of us to our hurt. But how strange again will be the day when it shall be altogether absent! Shakspeare somewhere speaks of a man " wearing his heart upon his sleeve, for daws to peck at ;"* but then, I suppose, we shall all wear our hearts outside, and

* *Othello*, act i. scene i.

not have a thought that we don't want everybody to see.

And then another thing is passing strange —the thinness of the veil between this world and·that, and yet the impossibility of ever seeing through it. I was thinking to-day of our dear good Archbishop.* Last week he was with us. Now, he is there. One who was part of our system—whose acts entered into my calculations—to whom I wrote and from whom I received letters, ay, and a very few days ago, —in an instant past the border and in that strange land. And we fancied as late as last month, that we should have him here with us, coming as if he were our father, and had had the bringing of us up all his life, and taking this little thin hand of yours, Margey, and asking after your welfare in a voice whose very tone made you better: so much was he

---

* Longley, who died October 28th, 1868. For some remarks on his enthronement and death see " Life of Dean Alford," 3rd edit. pp. 358, 419.—Ed.

mixed up with our every-day anticipations; and now we have to think of him as in that (to us) solemn world.

And do you remember, Jessie, when we stood, all four, round Edmund's dying bed,[*] with the sunset from the western sea filling the room with rosy light: and we watched till the dear features lost meaning and their lines stiffened, and then I pressed down the eyelids, and we left mamma with him, and we three went out bewildered, and sat down on the beach, and I said, Where is he now? I have it all before me—indeed that lovely bit of water-colour by Philip Mitchell which hangs there would bring it before me, could

---

[*] The Dean here describes the death of our last surviving son, Ambrose Oke, who died at Babbicombe, South Devon, August 31st, 1850, aged ten years. The water-coloured drawing, referred to, was bought by the Dean some years after: it is a sketch of the beach at Babbicombe, just below the "Carey Arms," where our dear child died. It hung on the left-hand side of the fire-place in the large drawing-room at the Deanery, close to the chair where the Dean always sat on Sunday evenings.—Ed.

I ever forget it. The sun had gone down, and had left in the lower sky a few lines of dull red, and under them the sea looked a pale ghastly blue (so it seemed to me)—and the sky above was clear, but as yet without a star. And there was not a sound, nor a breath, nor a ripple. All seemed to speak of a presence gone. He who had been about those rocks, and on that beach, and cleaving those waters—and now ?

Well, dear ones, it fairly beats us. And it's good for us to sit and think, that all our thoughts and hope about that unseen state, and those who are there, is simply and only faith, and nothing else. We *know* no more now, with our Gospel and our Bibles, about it and them, than the heathen know. I mean, none has described it to us: none has gone through the change, and come back to tell us how it was.

Simply faith: but then what a treasure of faith it is ! what a glorious thing is that resur-

rection from the dead! Whichever way I look
at it, it appears more wonderful and precious.
Now just think of it in this way.   Lay your
thoughts into mine, little Margey, as you are
laying your hand into mine.   I'm going to
take you through some dark and foul places,
but don't be afraid to go with me.

Think of the blessed Jesus, that last day of
his life.   You know Him by pictures.   But the
reality was hardly like the pictures.   His form
that day must have been very mean and vile
to contemplate.   Crushed, beaten, bruised, defiled : even his very look of heavenly meek-
ness (for we have seen such things) provoking
contempt, as of one broken in spirit and without courage.   Look on, my children : don't be
afraid of plain language, as long as it keeps
within the bounds of reverence; and those I
am sure you love that Blessed One too well
ever to transgress.   See Him cuffed, hustled,
knocked down by a brutal mob : hurried along
that street amidst the triumph of his enemies,

and at last—it is St. Peter's language, not
mine—nailed up to a log of wood—stripped
and shamed—an object which those who saw
beat upon their breasts and turned away.

Now, why am I dwelling on the least wel-
come parts of this description? Just to pro-
duce the effect I feel I am producing by this
little arm clinging tighter round my knee as I
speak. We cannot bear to think of this utter
shame of our dear Lord. Yes, but it's good
for us to think of it. The very darkness of the
cloud makes the sunshine brighter: the very
wailings of this dirge bring out the grand out-
burst of the chorus which follows.

Carry all that has been said in your minds
—then go on. The poor body—all paleness
and wounds—is taken down, is rained on by
many tears, is swathed by pious hands, is
carried to the dark chink in the rock. How
little thought Joseph and Nicodemus, as they
stooped and deposited it, flat and stiff and lifeless, in the recess of the cave, of anything un-

usual to follow! How little thought the holy
women, as they saw the stone rolled to the
door, what hands would roll it away!

But let us look again. It is morning, and
the risen sun glitters on the walls and battle-
ments of the city, and the garden trees wave
in the fresh awakened breeze, and the birds
are raising their spring chorus of praise. It is
the same cave—but open. Before it stoops the
Magdalene, weeping. Two youths sit within,
fairer than the sons of men, in pure white
robes. They ask her of her trouble. She
answers them. But what does she see? While
she is answering, she sees their countenances
alter, and she looks behind for the cause. He
himself stands by her. Incredulous at first
she knows the sound of the voice that pro-
nounces her name; she worships Him: she
runs to tell the disciples.

And from this moment onward, glory bursts
upon glory in the eyes of the wondering followers of Jesus. Where is the despised, abused

form now? Where the face defiled with shame and spitting? "Reach hither thine hand, and thrust it into my side:" what a glorious Body must have been unveiled when these words were uttered! "A spirit hath not flesh and bones, as ye see me have:" then his resurrection frame must have been compacted and built, even as this of ours, but we may conceive with what beauty and splendour. And in the moment when He rose in the pure summer noon from the top of Olivet, and the cloud received him out of their sight, what a vision must they have enjoyed of the glory of perfected humanity! What a streak of burning light on their memory must have been those forty days!

Thus much then we do know—thus much there is for faith to fasten upon. He went into that unknown world vile, despised, crushed, an object of aversion and scorn: he returned to show himself to our faith a perfect and glorified Body: the very type and flower of our Humanity. And the marvellous history as we read and rejoice, how full it is of human interest! that walk to Emmaus,—that challenge to Thomas,—that meal on the shore of Gennesaret! Had the risen Saviour flashed out a glorious apparition before their eyes, girt with attendant angels, and then been withdrawn—had they been rapt into a trance and beheld Him at the Father's right hand,—where would have been the assurance that now clings round our firesides and broods over our homesteads, that the Lord is risen?

And so, darlings, that is all man's eye has ever seen of the world on the other side—the reflection of its brightness from the glorified Body of Jesus. And we know, that as the firstfruits, so the harvest: as He is, so shall we be.

We have talked long enough, for I hear the church clock striking ten, and the servants will be ready for prayers.

Jessie, my love, look out " I know that my

Redeemer liveth:" you're in good voice to-day, and we shall all enjoy it. And, Margey, just run and tell Sampson to bring his violon-cello and his *Messiah:* and we'll end with "Worthy is the Lamb that was slain."

III.

"WELL, Margey, there's no fear of a sacred institution now. Last Sunday, preaching in London; the Sunday before helping poor Lacey in his terrible bereavement; and the Sunday before that, laid up with neuralgia. I think we've earned our homily to-night. And so we are to talk about Christmas, are we?"

"*Something* about Christmas, please, papa?"

"Oh, I see. That sly emphasis means, I suppose, not exactly *the main* subject, but some of those side ones which come with it. How is that, Margey?"

No answer, but a long look up at the wall.

"What is it? Oh, again I see; that lovely

picture of Andrea del Sarto's. Then I con-
clude I am to speak of the CHILD JESUS. The
tight-clinging squeeze tells me yes."

Well : First, then, let us put the Holy Child
in his right place. I'm going to take Margey's hint, and talk to you of the ways of
representing Him, and the beautiful visions
which art has given to our race, with Him
as their subject; but, before I do, let us see
how we ought to feel towards the Holy
Child.

Notice, then, dear ones, that the Holy Child
has no existence *now.* You can't pray to the
Child Jesus, because there is no such person.
It was simply a former state of Him, who now
reigns, perfect man, in his Father's glory.
He himself is for us what He is *now*, not what
He was once. If you or I pray to Jesus, we
can only pray to Him at the right hand of
God, exalted to be a Prince and a Saviour. To
pray to Him as a child, to pray to Him as on
the cross, is to pray to a mere thought, a mere

fiction. Such states of his are, you see, not objects of adoration for us; but they are most blessed objects of remembrance and of contemplation. Jesus as a child: behold one of the most beautiful objects on which the imagination of man can be fixed and employed. The only human child who has ever been sinless and spotless. All children are comparatively sinless; but this one was absolutely so. And then there are several things which give the contemplation even more of human interest. On the one hand, this child was born of a human mother. As a child is like its mother, so on the other hand it may be said, when the child is the principal thing to be thought of, that the mother is like the child. Something of the pure and holy and lovely countenance which this flower of children had, must his mother also have had, and must have thereby been the flower of maidens. And so, my darlings, you have at once the materials for the loveliest of all creations of human art.

Foremost among all the pictures in the world for beauty are those of Jesus, the flower of children, and Mary, the flower of maidens and mothers. Almost every one who has travelled has got together some favourite forms of this lovely group. On the walls of this room are no less than twelve, by different painters, collected during almost as many wanderings of mine.*

Before we speak of them we will mention another source of intense interest in the figure of the Child Jesus. In that infant form dwelt the Godhead; and the greatest and noblest of painters have ever borne this in mind; have made the baby-face, while all human in its

---

* Our drawing-rooms at the Deanery were full of remembrances of our foreign tours, photographs and prints of many well-known pictures by the great masters, a copy of one of Francia's paintings in the gallery of Prince Borghese, Rome ; a Virgin and Child, by Vivarini, bought at Venice in 1846; a Magdalene, by Sassoferrato, found at a picture-stall beneath the shadow of Milan Cathedral in 1841, and others too numerous to mention here.—ED.

tender beauty, yet, as well, something more than human. Look at that first and grandest of all such representations, the " Madonna del Sisto " of Rafael. Yes, Margey, there it is, that large splendid lithograph, which I shall never repent having bought when I was at Dresden, where the immortal picture is. Look at the glorious infant. Does not the very Godhead burst from those lips and those wonderful large eyes? No other painter ever imagined a face like that. If you saw such a child as that, you would almost say, as the beloved disciple said to Peter, " It is the Lord ! " And the mother, too. How remarkable it is that Rafael, living among all the error and nonsense which Rome has accumulated round her, should have painted such an exquisitely simple form, a form and face saying nothing but, " Behold the handmaid of the Lord ! " And yet there is no want of grandeur : she knows that she has been the mother of the Divine Child—every feature speaks of it

—and before the mighty truth she is subdued
and humble. No such woman has ever before
been painted, nor since. The hour which con-
ceived the Madonna del Sisto was the noontide
point of human genius.

Now let me point out, darlings, one thing in
the representations of the Holy Child. He is
all pure—without sin : in his blessed body
there is no room for shame. And all the
greatest painters, in their greatest pictures,
have taken account of this. As our first
parents in Eden, so the Infant Jesus is wholly
naked. There is no reason, when we represent
the ideal of Him, for veiling that human form
which in itself, and apart from sin, is very
good. Bear this in mind, for we shall have
to recur to it when we speak of the group
enlarged.

Now next, for we are yet concerned only
with the mother and child, look up at that exquisite but quaint idea of Francisco Francia's.
The mother stands in a field of flowers, looking

down with admiration and reverence on her Child, who is lying on the green sward, looking up at her with a face beaming with love and heavenly benignity. The picture will show what variety there was in the conceptions of the beautiful group; and you will be glad to hear that between this painter and the young Rafael there was the closest and most admiring friendship.

Another of his there is on the wall—a lovely Madonna and Child, from the gallery of Prince Borghese at Rome. It is almost singular among those which we have here, for it is in colours. I saw a water-colour painter making that copy in the gallery, and I bought it. It is Francia's very best. Notice the exquisite colours, the deep crimson of the Virgin's inner vest, overlaid by the rich blue of the outer robe which envelopes her head, and flows down to her feet. Then remark the heavenly clearness of her complexion shared by the blessed Child who sits in bright nakedness on her lap,

resting his hand in hers, while her other clasps
Him round in closest affection. See also how
Francia falls beneath Rafael in the full expression of both human and divine. There were
thirty-four years difference between them :
representing thirty-four years, as you will see
when you come to read the history of art, of
great and rapid progress.

But now look up, little women, and gaze for
awhile at that quaint picture, with gold ground
and heavy ornamented frame. There sits a
stiff Virgin, certainly without any claim to
beauty, with a queer indented glory round her
head, and her hands up, as if in adoration ;
and on her knees a perfectly lovely Child,
asleep, painted with the most elaborate care,
almost like enamel on porcelain. That little
picture I bought at Venice in 1846, and it
was certified to me as being by Bartolomeo
Vivarini, *i.e.* as painted about 1480. It may
serve to you as a specimen of the hard, early
style of that Venetian School, not without

great dawning promise of tenderness, and the good symptom of diligent and religious carefulness.

If I were to take you back, Margey, to the Madonnas of Rafael, we might spend all night in descriptions; but there you have several of them, and in my box of photographs many more. It is only wonderful how he devised so many attitudes and circumstances under which to place them—all beautiful, though not all equally beautiful. One of the most lovely is that known as the " Madonna di Gran Duca," now in the Royal Palace at Florence. It seems as if the painter had poured his whole soul of tenderness into this group. The Divine Child is sitting on his mother's hand, the lines of his figure almost coinciding with those of hers. Contrary to Rafael's usual practice, He is wound about the breast with a band of full drapery. She looks down on Him with a countenance full of reverence and love, while with her other hand she steadies his body, holding it under the arm. His little hand rests lovingly on her bosom.

And now, my darlings, I think we had better reserve the larger groups of the Holy Family, of which there are some very lovely examples, for another of our evenings.

I see it is getting late, and the Laceys, poor things, who have just now small comfort at home, are coming in to prayers, to help us sing, "Unto us a Child is born."

IV

YES, Jessie, as you observe, that must have been a very blessed home at Nazareth. How little we think of those thirty years! Thirty out of thirty-three—why, according to what we were learning yesterday in fractions, that is ten elevenths of our Lord's whole life on earth, isn't it? Certainly the three that remained were much more important, because they contained his teaching and his death : but it seems out of all reason, to entirely neglect the thirty which preceded.

The great Christian painters have not done so. They have loved to represent the Holy Family as they were at Nazareth when Jesus was a child, and when John, who you know was six months older, may have been his playmate and companion. I am going to describe to you several of their representations of this beautiful group. Yes, Margey, I feel the squeeze. What? Only a whisper? Well, never mind, provided I can hear.

Yes, no doubt you are right, little one: Margey asks, why the brothers and sisters are not painted too? Well, little lass, I am afraid the old Catholic people did not think much of them. They would have it, you see, that the mother of our Lord had none other but Him, and they managed to make out that those whom He called his brothers and sisters were in reality his cousins. But if they were, then their own mother was living, and it seems unaccountable if she were, that they should always be mentioned with their aunt, and not with their mother. There are several other reasons too, against this curious evasion of the plain sense of Scripture: one of which is that, at a time after the Twelve Apostles had been chosen, among whom according to these people are two of our Lord's brethren (or cousins), it is said "For neither did his brethren believe in Him." Your question, little woman, has just anticipated a remark I was going to make before the end of the evening; that the picture of the family at Nazareth has yet to be painted, and a very beautiful one it would be. Those sisters of our Lord, what were they? How did they think of Him? Of this we know nothing.

There doubtless was another reason why no more than our Lord and St. John among the children are represented: and that is, that they two are the only persons full of deep meaning to us Christians. The rest may have been very holy and good: but they are nothing to us in themselves. They have no part in the history of Redemption. Whereas John was the great Forerunner, the pointer to our Lord as the Lamb of God. And thus he finds his place always in the symbolism of the picture.

So we will now speak of those beautiful groups which we have, and will leave off complaining that we have not more. We begin of course with Rafael. And the first which I will mention, because one of the most infantine, is that favourite of yours, little Margey, hanging on the left side of the fireplace. The Child Jesus is lying asleep on a piece of drapery, and his mother is lifting a veil from off Him and showing Him to the little St. John, who kneels in childish admiration, his lips apart and his hands joined. Certainly there hardly can be anything more peaceful and beautiful. The sleeping Infant is the same Divine Child as we have opposite in the Madonna del Sisto, with the wondrous face subdued and softened by the human infirmity of sleep. His figure is the very essence of repose—"full of rest from head to feet." Still there is one thing about this sweet picture which interferes with its purity and simplicity. Though the Child is the same as in Rafael's greatest picture, the Mother is not. She is crowned, and thereby a concession is made to the superstition of Rome which degrades her from her dignity of humility, and crowns her as a heathen goddess, calling her the Queen of Heaven. In consequence of this peculiarity, the picture is known as the " Diadem." It is in the gallery of the Louvre at Paris.

Now next observe that other hanging on the right of the fire. Here we have the blessed Mother sitting in a lovely garden with flowers springing around; at her right knee stands the Holy Child, in purest nakedness: his left arm rests lovingly on her left hand, which lies in her lap: His right hand touches her knee, while her right hand embraces Him under the right shoulder. He is looking up at her with a look of unutterable love, while she casts down her full eyes on Him in love mingled with reverence. Next to the Madonna del Sisto, this seems to me the most beautiful figure of the Virgin; it is so simple in faith and adoration, yet so dignified, and thoroughly worthy of her exalted relation to God manifest in the flesh. The little St. John kneels on the ground on one knee, but it always has appeared to me in rather a constrained attitude. As we now have the picture there is something not altogether intelligible about his face; but the painting has been done over and over again since Rafael's time, and probably the face has been damaged by some inferior hand. Notice in this, as in most other groups of the kind, that whereas the sinless Child is without covering, the other, one of us, is girt with drapery. This picture, commonly known as "La Belle Jardinière," is also in the Louvre at Paris.

The third of Rafael's groups which I will notice is given in this large photograph from the picture at Florence. I will read you a description of it, which I wrote five years ago :—*

"One picture, in the Tribune of the Uffizi Gallery, especially struck me. The *Madonna del Cardellino* (our Lady of the Goldfinch) is one of the loveliest works of Rafael. The Virgin is sitting on a rock, in a flowery meadow. Behind are the usual light and feathery trees, growing on the bank of a stream, which passes off to the left in a rocky bend, and is crossed by a bridge of a single arch. To the right the opposite bank slopes upward in a gentle glade, across which is a village backed by two distant mountain-peaks.

"In front of the sitting matronly figure of the Virgin are the holy children, our Lord and the Baptist, one on either side of her right knee. She has been reading, and the approach of St. John has caused her to look

* See the Dean's "Letters from Abroad," chap. vii. p. 205, second edition.

off her book (which is open in her left hand) at the new comer, which she does with a look of holy love and gentleness, at the same time caressingly drawing him to her with her right hand, which touches his little body under the right arm. In both hands, which rest across the Virgin's knee, he holds a captive goldfinch, which he has brought with childish glee as an offering to the Holy Child. The Infant Jesus, standing between his mother's knees, with one foot placed on her foot, and her hand with the open book close above his shoulder, regards the Baptist with an upward look of gentle solemnity, at the same time that He holds His bent hand over the head of the bird.

"So much for mere description. The inner feeling of the picture, the motive which has prompted it, has surely hardly ever been surpassed. The blessed Virgin, in casting her arm around the infant St. John, looks down on him with a holy complacency for the testimony which he is to bear to her Son. Notice the human boyish glee with which the Baptist presents the captured goldfinch, and, on the other hand, the divine look, even of majesty and creative love, with which the Infant Jesus, laying his hand on the head of the bird, half reproves St. John, as it were saying, 'Love them, and hurt them not.' Notice too the unfrightened calm of the bird itself, passive under the hand of its loving Creator. All these are features of the very highest power of human art. Again, in accompaniments, all is as it should be. The Virgin, modestly and beautifully draped: St. John, girt about the loins, not only in accord with his well-known prophetic costume, but also as partaking of sinful humanity, and therefore needing such cincture: the Child Redeemer, with a slight cincture, just to suggest motherly care, but not over the part usually concealed, as indeed it never ought to be, seeing that in Him was no sin, and that it is this spotless purity which is ever the leading idea in representations of Him as an infant. Notice too his foot, beautifully resting on that of his mother : the unity between them being thus wonderfully, though slightly, kept up. Her eye has just been dwelling on the book of the Prophecies open in her hand; and thus the spectator's thought is ruled in accordance with the high mission of the Holy One of God, and thrown forward into the grand and blessed future. It is a holy and wonderful picture : I had not seen any in Italy which had struck or refreshed me more."

Rafael painted many, many more such groups, which I should like to show my darlings, where they shine in all their heavenly colours, some day. But we must leave the Prince of Painters now, and notice two or three other groups of the Holy Family, which are on our own walls.

And first among them comes one which I have hung in the most conspicuous place in the library, because it was intimately connected with my own childhood. In my father's study, over the fire, was that Holy Family, by Andrea del Sarto.* Day by day, at my lessons, I used to gaze on that lovely figure of the Divine Child, standing in the front of the picture; his right knee bent on his mother's lap, who is half sitting, half-kneeling on the ground, his left hand passing across his body, and laid lovingly on the drapery of her bosom. To his left, and to the right of the picture, St. Elizabeth, a venerable matron, holds under the left arm with one hand, and on the right shoulder with the other, St. John, a lad of some seven or eight years, girt with his leathern girdle; his right hand, which passes under his mother's arm, is lifted in a pointing attitude, as bespeaking attention to his proclamation of the Saviour. Behind the group, on the left of the picture, is an angel, with an instrument of music; and behind him

* This picture was hung over the mantel-piece in his own library at the Deanery. See " Life," p. 479.—ED.

again, not very clearly expressed, the hand and back of the head of another. The faces are not of the high order of Rafael's, but still very beautiful, and evidently aiming at the same effect.

This pointing out of the Lord by his infant Forerunner is very commonly indicated in groups of the Holy Family. You will see it very marked in that lovely picture by Scarsellino di Ferrara on the opposite wall. The two children, both naked, are lying on a green sward with trees in the background. Both have been asleep, but St. John has awakened, and half sits up, pointing at the Saviour. The expression of his round boyish face is charming: his full eyes dilate with the great message of which his newly awakened consciousness is full: while the Divine Child lies peacefully slumbering by him, his left arm hanging relaxed over the advanced knee of the elder child. The original is at Munich.

One more picture, Margey, and those heavy

eyelids shall seek their pillow. But that one I cannot possibly pass over, if it were merely that you may have something by which to vindicate one of the greatest of painters when you hear him abused as he sometimes deserves. Ah, I see both the faces, and mamma's too, looking at the opposite wall. Yes, it is that group of four children by Rubens, that I mean. Hardly a purer and lovelier picture of the kind exists. It is allegory mingled with matter of fact. The little naked figures on the ground are four. First our Lord on the right, sitting with his left side towards us. The clearness and purity of flesh painting and colour of his body have surely never been surpassed. Facing Him, but with his back to us, St. John is turning to give heed to something that the Lord is telling him. His face is rapt in earnest attention, and his eyes look into the Face opposite with wonder and deepest love : the Lord is evidently laying forth to him that which he must proclaim and suffer for his sake. With his right hand He affectionately caresses St. John's cheek, while He regards him with a look of superior wisdom and love. Between the two are seen the face and shoulders of a little girl, known to be such by the arrangement of the hair. On the left, an infant angel is bringing a lamb, over which St. John casts his left arm, his right resting on the right thigh of our Lord. The meaning of this is plain. The little girl is the spouse, the Church. She holds in her hand a bunch of grapes, the well-known emblem of the Holy Communion. The lamb serves to indicate the whole deep mystery of the atoning Lamb of God, which, as the Baptist's special message and proclamation, is being declared by the Lord, and as signified by the arm laid over the lamb, recognised by the Forerunner.

There seem to be two originals of this wonderful picture. The German lithograph on the wall I bought at Berlin in 1857, after seeing the painting in the gallery there; and another was exhibited the other day at Leeds by Sir John Ramsden. This latter was engraved in the *Illustrated London News* of October 31, last year.* The group is, to my mind, not improved in it by an enormous canopy of fruit and flowers by Seghen, which does not exist in the Berlin original.

Well, darlings, I really have run on with this interesting theme, till we are all well tired. So we will not sing to-night, but reserve ourselves for a grand chorus, with a new subject, next Sunday.

* 1868.

V

THE next thing about the Child Jesus ? Well, I suppose you don't mean his being taken into Egypt, or presented in the Temple, or adored by the Magi,—because all these are hardly things about *Him*, but, taking his human course as that in which He is manifested, these are rather circumstances happening to Him than events with which He was Himself concerned.

The next thing about Him—well, Margey, what shall we say ? I know what occurs to *me.* I wonder whether we are agreed in this. You mean that staying behind in the Temple : is that it ?

Well, then, of that let us speak. It is an incident full of interest in many ways. I need not go over all the particulars of it; but will only just bring out one or two points which you, darlings, will feel, and are not commonly dwelt on.

The first of these is the marvellous way in which the Holy Child was trusted. When the caravan was made up on leaving Jerusalem, his parents do not seem to have given themselves one anxious thought about Him. The question, where He was, never arose till after the day's journey. "Then came still evening on," of which "the ancient poetess singeth," that it brings the child to the mother.* We may imagine the tent set up, and the evening meal prepared, as Eastern travellers describe it to this day. How many there were within it we cannot say. We believe that there were many brothers and sisters of the Holy Child; but as all were necessarily younger than Himself, none of them would

* Sappho. See "Chapters on the Poets of Ancient Greece," by Henry Alford, p. 75.

accompany their parents on this journey: unless indeed there were an infant which could not be separated from its mother. Perhaps there were others, too, who were nearly related to the Holy Family: but from what follows it would appear that none occupied the same tent with them. So that perhaps Joseph and Mary were alone. The sun dropped burning below the western plain, or dipped under the sea-line, if they were in the hill country: but still He came not. So completely did they trust Him that they waited till the great stars blazed out above: still He came not. Then they went out into the neighbouring tents and made inquiry. "He is sure to be in one of them," they said one to another: "for all love Him as their own, and He is fond of talking to them in his strange gentle way about great and holy things, so that He and they both take no note of time when so employed."

One tent after another was searched, but He

appeared not. Even then we may well suppose that distrust was not allowed to enter their minds. Not only had they both knowledge of his wonderful character, not only had his mother laid up in her heart all the mysterious matters regarding his birth, and the prophecies which were uttered respecting Him, but in all this experience of his childhood, and his boyhood, He had never once deceived, never once disobeyed them. He might have said to them as He said to his enemies twenty years after, " Which of you convinceth me of sin ? "

This being so, it was natural that sorrow should have filled their hearts. You, darlings, hardly yet know how sharp a pang is the first possibility of suspicion of one whom you have deeply loved : when the heart refuses to form to itself the darkening thought that is waiting to take shape, and we say " it must be right," when we nearly half feel that something is grievously wrong.

Some such pang may have rent the breasts of Joseph and the blessed mother that lonely night as they lay sleepless in the tent. We may well conceive that again and again, separate and together, they joined in prayer to Him who had given the precious gift that it might be theirs again.

Oh what a feeling this is, the sense that *one is lost*—not lost, as those that are gone from us are lost—but lost, perhaps never to be found!

I remember, when I was at Quebec Chapel, in London, a very dear friend * preached one of my Friday Lent sermons on the text, "Whosoever will come after me, let him deny himself, and take up his cross and follow me." He insisted on the fact, that every follower of the Blessed Lord has a cross to bear—he told us of the frequently sudden coming upon this cross in our way—when we least were

* The late Rev. J. H. Gurney, Rector of St. Mary's, Bryanston Square.—ED.

thinking of it, we turned a corner, and there it lay, to be taken up and borne.

The service ended and we came into the vestry. There stood his eldest girl, pale and breathless—" Oh, papa, baby is lost in the Park! She strayed away from nurse, and can't be found." " I little thought," said he, turning to me, " that at the first corner I should find the cross ! "

And now Joseph and Mary had found their cross. " Sorrowing "—it is her own word— sorrowing they sought Him—turning back from the tents of their companions at the early dawn.

A fanciful Roman Catholic saint, Bonaventura,* in whose meditations are some beautiful thoughts amidst much that is nonsensical, suggests that the blessed mother may have

* That good Bishop of Albano, at whose funeral, A.D. 1274, the Pope, and cardinals, and five hundred bishops attended, and also the Patriarchs of Constantinople and Antioch.—ED.

thought that the same Father who had given Him to her may have again taken Him away. But this, in the face of the prophecies to Joseph and to herself, could hardly be. Much rather might she perhaps have suspected, that that mysterious mission, of which those prophecies had faintly hinted, may have begun: that He may have betaken Himself to the desert, there to await the time of his manifestation to Israel.

At all events, Jerusalem was the point at which their search must begin. And now from one wonder we come to another. We have seen how implicitly they trusted the Holy Child. But if we were to assume from this that all his heart and purposes lay open to them, we should be wrong. And this part of the history reveals to us in a wonderful manner the human character of our Lord. We read on one occasion in St. John, that He "did not commit himself unto" the Jews who came up unto the feast at Jerusalem: and even so had been his practice through his younger years. He was a pure, unsinning child—an affectionate and obedient child— but at the same time a reserved child. Words of comfort, words of peace, words of tenderness, He must always have had for those about Him : but they were only such as He pleased to utter : they came from a depth within which no one knew : they were parts of a store which human thought never measured.

How do we know this on the present occasion ? Simply by this circumstance : that they, who had been with Him now twelve years, never for three days thought of looking for Him in the place to which all his boyish enthusiasm tended. The house of his heavenly Father had never been evident as an object of his love or admiration. Long, long years at Nazareth had the desire been growing in Him, to dwell in the house of the Lord, and behold the fair beauty of the Temple :

long had He yearned after the converse of holy men whose time and toil were spent over the Law of God: but of this not a word had escaped Him.

And this interests us not only as being a trait of Him, but as so completely identifying Him with the boy of all times and countries. Nothing is more common than for the ruling passion of a boy's soul to be thus kept in reserve—father, and even mother, not having a dream of it—till some day it leaps forth full-grown, and amazes all who behold.

This is often so with a boy's plans about his destination in life : but, it is true, not so often with regard to favourite habits and haunts. It conveys to us a stronger idea of the reserve which must have been the characteristic of the Holy Child : and serves to mark that character as no common one, even in its quiet and private manifestation. This mixture of affectionate tenderness and habitual reserve can only exist in the great and few spirits of the very highest order : and that these did coexist in the Child Jesus, at once points Him out as unlike the common run of the sons of men.

But, even allowing for all this reserve, it does seem passing strange that Joseph and Mary never during those three days seem to have thought of the Temple. We might ask, Where did they seek Him all that time? and what put into their thoughts at last to try the place where He was? They perhaps went to the family in whose house they had celebrated the Passover. Was it the same as that pointed out by the Lord twenty years after when he said, " Go into the city to such a man and say, The Master saith, I will celebrate the Passover at thine house with my disciples "? There they might seek in vain ; and then perhaps they branched out among those who had been their fellows in the sacred feast, but equally in vain. One day passed, and another, and another : and there was doubtless the hope deferred, that maketh the heart sick : more and more anxiety, at last fading off into the blankness of despair.

At last, we may well imagine, a rumour reaches them—that it was so long in reaching them may at once serve to refute the legends about the early connection of his mother with the temple of the High Priest—a rumour that a wonderful boy was in the Temple, sitting in the midst of the doctors, hearing them and asking them questions : and that all were astonished at his understanding and his answers. Who could this be but their lost one? That understanding — those answers — how often had they astonished those parents at home! How often, we may well imagine, had they said one to another, " What rabbi could have answered more wisely ? "

We all know how many wise things are said by children. You, Jessie and Margey, in your time, have both of you been utterers of wise sayings, such sayings as you will never, never utter again. Jeremy Taylor* says that children suffer, when they are in pain or sorrow, by direct pressure, as a pillar supports a weight. The idea is very beautiful; and we might expand it further, and say that when children think, they think by direct contact with truth, without those side views and obscuring compensations which disturb their thoughts in after years. So that we sometimes have more living truth, direct from God's Spirit, in the saying of a child, than in ten mature verdicts of grown-up men.

And if this be so with you and others who grow up in imperfection and under all the beclouding influences of selfish temper and departures from truth, how must it have been with Him who never sinned?

When He asked a question of the doctors of the law, it was not for display, it was not from

---

* Who was said to have devotion for a cloister, learning for a university, and wit for a virtuoso college. He died in Ireland, 1667, bishop of Down and Connor.—Ed.

idle curiosity, it was not for the love of victory in argument : it was the earnest reaching forth after truth of a soul which basked in truth : the inquirer courted no flattery, deserved no rebuke, stirred no jealousy, overstepped no modesty ; it was as if Truth herself, radiant as the bow of God, had stepped unclothed from her veil, and won all hearts by her smile.

Let us look somewhat more, darlings, at that wondrous assembly. Yes, Margey, there it hangs * before us, but without its glorious colour, as Holman Hunt gave it forth from the year's study of his earnest soul. I wish you could have seen the picture, all aglow with those wonderful hues—somewhat, perhaps, too rainbow-like and shifty in gleams, but yet no tint without meaning, and all conspiring to one of the most glorious of effects.

It was some such assembly as the painter has there represented. The grand old rabbi,

* The Dean wrote this description with the engraving placed by his side on a chair.—ED.

whose winters mounted to a century, their snowy marks on his scanty beard, and their film over his sightless eyes—how he clasps the great scroll of the law, the study of his life, and the fathomless well of his ripened wisdom! The aged compeer at his side laying his hand on his arm, is setting forth to him the reason why the wise and holy talk of the young peasant from Galilee has of a sudden ceased. And next to him is a young teacher, his face full of intelligence, his brow contracted with anxious thought as over some answer from which the very soul of righteousness had looked forth, or over some question which the collective wisdom of rabbidom was all too poor to furnish with a reply. And so we pass on, to some faces which look secular, and even some which seem, but probably are not, void of meaning, till our eyes reach the right-hand, or principal group of the picture.

And here what shall we say ? I know that tastes differ among us on this group ; I know also that my own feeling has not been always the same about it : but I also feel that the artist had immense difficulties to contend with, and that he has surmounted them not by pandering to conventionality, but by patiently studying and then idealizing nature.

Let us take them in inverse order and importance.

The figure and expression of Joseph are, to me, faultless. There is no assumption of importance in them, as neither ought there to be : but the great joy of having found Him who had been lost is mingled with a serene satisfaction at the place and employ in which He has been found : and thus this manly peaceful face sets, as it were, the tone of the group.

Of the Blessed Mother more must be said : more which may call, and which may be called, in question.

The expression is as of one earnestly and passionately pleading ; as we might imagine her to have done, had we not been told expressly that she did not. The account given in St. Luke certainly does not lead us to think that she thus earnestly and closely whispered in the ear of her Son. There is in that narrative a majesty of motherhood, which I fail to discover here. Perhaps it may be said, that the artist has altogether *translated* the narrative into detail; that the saying in St. Luke is that to which all her dealing with Him amounted, rather than any one portion of it; that we can hardly imagine the joy of finding, the intense interest in the situation, the desire to win Him back again—all venting themselves in those few and balanced words; and that though the Evangelist is faithful to the summary of fact, the artist has seized on one of the expressions of nature of which that summary was made up. It may be so. Painting, we know, is tied to a moment, and must give an outward act done. History is tied merely to truth, and truth may be the total of a great many acts.

But perhaps all this is too hard for my little women ; and at any rate you will see in the attitude and expression of the Blessed Mother what, if it represent not the whole sacred narrative, must have been gone through before that whole was attained.

But now we come to speak of the figure of the Holy Child Himself. And I hardly know how to praise this too highly. It seems to me to have just that mingled look of human boy-hood, and divine yearning for higher things than human, which we should expect, but look for in vain, in any representation of the youth-ful Jesus. It is found in the Infant of the Madonna del Sisto, and as has been said, in one or two other of Rafael's ; and, as far as I know, in those only. That the earnest desire to be "among his Father's matters" is here somewhat prominent, is hardly to be blamed : but none can say that the rising resolution to check that desire, and to go down to Nazareth and be obedient to them, is not also abundantly expressed. There is one little incident of the Lord's posture which has always struck me as very beautiful ; the playing of the right hand with the buckle of the band. It exactly expresses the meeting of two currents of feeling. One can see in this as in the face, the truant interest in the disputation of the doctors, wavering before the strong return of self-denying duty ; while, at the same time, there look out wonderfully from the eyes the thoughts that come from otherwhere than this our earth.

Of the accessories of the picture it is after this hardly worth while to speak. According to the artist it is evidently full day. Workmen are shaping a stone outside. A beggar is laid at the gate to ask alms of them that came in. Now I had in my own mind always imagined it evening ; " After three days they found Him in the Temple." Whether the doctors had the custom of sitting on there till the evening, I am not sufficiently acquainted with Jewish practices to be able to say; but the " after three days " seems to point this way. Perhaps the wonderful understanding and answers of the Divine Boy may have kept the dignified conclave beyond its ordinary time of sitting.

Leaving the picture, we may finish by saying, that it is full of interest to think what the Lord may have seen on that his youthful visit which led to his words and acts years further on.

Then He may have noted with boyish indignation the unhallowed practice of buying and selling in the house of prayer, which twenty years after prompted one of his first, and again one of his last acts of summary vindication of the holy law of his Father. There he may have noticed the Pharisee standing and uttering his self-satisfied prayers, and the poor publican standing afar off and beating his breast in contrition.

The conclusion of the beautiful story is as

teaching as all else is. " He went down to Nazareth with them, and was subject unto them." His holy burning zeal was curbed and repressed. " Even Christ pleased not Himself." The outbreak of enthusiasm had been an infirmity, not a sin : and now the well-balanced spirit again righted and strengthened itself: the course, which begun by becoming obedient, returned into its proper channel.

And to us the wonderful part of such obedience is, to think of its duration. Not for a month, not for a year, but for eighteen long years was he subject unto them. He became " the carpenter." Doubtless many a house in Nazareth witnessed his humble subjection to his reputed father's trade. One of the earliest fathers, Justin Martyr, tells us that He made ploughs and yokes.

There was some years ago a striking picture of Herbert's, of our Lord, as a lad of fifteen or sixteen, in Joseph's workshop. He is no longer a beautiful Child, but a thin workman youth. He is, as is the custom with workmen there, naked, except for a cincture round the loins ; and the painter's effort has been rather to pourtray the bodily fatigue incident to his life of unflinching obedience, combined with the higher purpose manifested in his holy looks. Perhaps the painful part of the picture is carried too far, as is the practice with the school to which Herbert attached himself: but it was a picture which entered that night into one's waking thoughts ; and that is, at least in my little court of criticism within, no small praise.

There is another picture of Herbert's, belonging to this his period of boyish obedience : I am not sure whether or not to a time before the incident we have been dealing with to-night. The Child Jesus is passing along Joseph's shop, when suddenly his eye falls on two chips of wood, accidentally fallen one over the other in the form of a cross. He pauses,

looking down on them with a solemn air.
His Blessed Mother stands in the doorway,
contemplating Him with an air of conscious
earnestness.

And even so ends our present narrative.
"But his mother kept all these things, and
pondered them in her heart."

Of that we shall have more to say another
time.

Now for prayers; and we will try "The
Lord whom ye seek shall suddenly come to
his temple : " and the grand chorus, "And He
shall purify," which follows.

VI.

SO you want me to say something about that first miracle in Cana of Galilee. Well, it is a scene my mind often dwells upon. I love to think of our Blessed Lord among the common incidents of life. You two darlings have as yet young fresh hearts, and are easily moved: but by the time you have been fifty years knocking about in this heartless world, you will find the gristle got very hard and callous about most of the places that are soft now, and it will take something unusual to stir that salt fountain inside. And it's strange what kind of things do it, when nothing else will. There's that sweetest bit in all old heathen poetry, that I tried to read to you, Margey, the other day, where the great grand warrior lifts his little soft child, having taken off the nodding plume which made it shrink back into its nurse's arms, and kisses it, and prays for it.* Well, I used at school to think that very commonplace if not almost childish: but I never can read it now without a hot tear brimming over. And when anything at all like it happens in real life: when incongruities from two opposite sides of humanity meet on humanity's common ground, the same takes place. Other incidents lose their freshness, but these never.

And one of these is, our Blessed Lord at a marriage. That bridegroom and bride, those cousins and gossips, I suppose they were just common ordinary people. I suppose the bridegroom had arrayed himself in his best, and the bride had been decked by her mother: one of the pretty maidens of the neighbourhood of Nazareth, of some twelve or thirteen

* Hector's parting with his child: Homer's "Iliad," book iv.—Ed.

years—nay, Margey, don't start and look so surprised, for it was so, and is now in those parts—and there was all the talk that there always is at such times—don't let us be afraid of imagining it, for the more real are our ideas of all about Him the nearer shall we approach to Himself. The looks of the bridegroom—this for the damsels—those of the bride and her maids, this for the men; and perchance the last news from the petty war the other side Jordan, or the damage done by last week's hail, or the prospect of this year's vintage: or perhaps some talk of lucky or unlucky omens that day, and argument, backed by sayings quoted from learned Rabbis: and appeal to the local scribe, or priest, or Pharisee, ready in traditionary lore. All this would no doubt issue in that clatter of tongues, which of all things is the busiest and the most absurd to listen to from without, one's heart not in it, one's ear unclaimed by any sound of them all.

So far, might be an ordinary meeting of
guests at any wedding on earth : or at least
it might, changing changeables : but here at
once we come to something which brings the
heart into the mouth.   HE, there?   And how
would He comport himself there?   We are
apt to think of Him as never unbending.
But it could hardly have been so.   Thirty
years they all had known Him.   Perhaps
none beyond the inmost circle of his family,
perhaps none but she whose bosom had pil-
lowed his infant head, knew of the wonders
of the birth at Bethlehem.   What did they all
think of Him?   At the end of boyhood, we
left Him increasing in wisdom as in age, and
in favour with God and man : at his baptism,
his relative, not knowing his loftier character, says, " I have need to be baptized of
thee."   These are the two testimonies which
come down to us, only these, from those thirty
years.   From which however we may gather
much.   That first one and its attendant cir-

cumstances are full of interest, as we saw last month. Let us remember what we said there. We said that it must be a strange boyish character which on the one hand was so thoroughly trusted that his parents should go a day's journey in ignorance where he was in the company: and which on the other hand had so thoroughly kept under its leading enthusiasm, that they never thought of seeking Him where He wondered that they had not at once sought Him. And we said on the former of these, thoroughly trusted is warmly loved: trust comes through love. And thoroughly trusted had another side, as we also saw then. "Where is He?" said Joseph. "Safe among our kinsfolk or acquaintance," replied the blessed mother: "they all love Him like their own child."

Now, from what we then said let us go on. He increased in wisdom—He increased in favour,—as He increased in age. So that He must have been known, in that privacy of neighbourhood life, as a very remarkable and a very popular person. That unique character among the sons of men, half masculine majesty, half feminine tenderness, must have appeared long before it was manifested to Israel. Doubtless, many had come to Him for comfort. Doubtless even as a child He must have had his little tender-hearted followers. He must have wiped away many a tear, long before He put on his harness for the victory which shall wipe away all tears.

"I have need to be baptized of thee." Think who speaks the words : one who was filled with the Holy Ghost from his mother's womb : one who was a pattern of holy purity and self-denial. The words meant, "So faultless art thou, so gentle, so wholly unneeding anything that my baptism represents or requires, that Thou oughtest to be the baptizer, and I, who day by day need penitence and need purifying, ought to be the receiver of the ordinance at Thine hand."

Such then was the wedding Guest: the beloved of all, the trusted by all, the counsellor the comforter. Yet, from what I said we must needs suppose that there was in the blessed Jesus great reserve: probably something like habitual silence: certainly considerable jealousy of the prying eye or the interfering voice, as regarded any unfoldings of his own most mysterious course, opening, opening, now year by year.

We read respecting the wedding, " And the mother of Jesus was there." It would seem as if she were hardly a guest, but one who had a right to be present. The same also appears from her ordering the servants afterwards.

The mother. No one else? It is commonly, and I suppose rightly thought that Joseph was ere this gathered to his fathers. It is a strange thing to think of the holy youth having stood by a death-bed, before his power over death was manifested. Could we but lift the veil, and see how He mourned for his just and noble parent—what wonderful words of comfort He addressed to his mother and brethren and sisters—how perhaps He went up the hill to pray, and continued all night communing with that Father who could not be taken from Him.

One of Margey's little secret questions. Well, little one, it may be so : I cannot say, but I have sometimes thought it.

Margey asks whether we shall ever be able to inquire about those years. Why not? Why should we not sit round Him on the flowery banks of the river of the water of life, and hear Him tell of the thoughts and incidents of that lovely childhood? Why should not He say to us of the Gospel as He once said of the Law, "It was said to them of old time," "He grew in wisdom as in age : " "but I say unto you that on such a day . . . ." Ah, we cannot fill up one such narrative now.

But to return to the wedding feast. Jesus

was bidden, and his disciples. So that by this time He had around Him a body of recognised followers: those I suppose who had been spoken of as called by Him in the chapter before.

And so the meal went on: but what followed was somewhat strange. The wine ran short. On such occasions generally the supply is abundant. What was the cause does not appear. It may have been the poverty of the family, or an unexpected accession of guests, but so it was. And apparently the defect was not quite unexpected. There must surely have been some previous conversation on the subject, or the mother of Jesus could hardly have said what she did. We know how she watched Him: how she laid up all things concerning Him in her heart. It may well have been that there had been in the house at Nazareth some foreboding of such failure, and fear of the giver of the feast being put to shame: and that some general remark of his, unnoticed

G

perhaps by others, may have set her motherly heart beating high with expectation. For we are utterly unable to comprehend the blessed and intimate relation of a mother, and such a mother, towards such a Child—the eagerness for his manifestation, now so long delayed—the impatience of his seeming reserve and want of ambition, with all these years slipping away beneath her eye. So she comes forward in the character of his prompter and patroness —with somewhat of a woman's and a mother's desire to show off the high dignity of her Son : she said unto Him, " They have no wine."

It has been asked, What did she expect Him to do? because we have it from St. John that this was his first miracle, and therefore she could not have looked forward to anything so out of the common course from Him. But I think those who ask this question know but little of the very close relation of which we have just been speaking. She had long and narrowly watched his thoughts and words.

The idea, of commanding the powers of nature by a power above nature, had been ripening in his mind perhaps for months. If He spoke of such matters at all, especially if there had taken place anything like the conversation we have mentioned, the slightest hint dropped by Him could not fail to speak meaning to the watchful mother's heart.

So she thought and thought, and bided her time. And it was with her, as it often is with us: a bold idea, at first entertained with reluctance, becomes by degrees familiar to the fore-castings of action: we fancy ourselves uttering it, we shape the words in which it is to take form, we rehearse it again and again in our minds: and when the anticipated moment arrives we act, not the modest part which was at first suggested, but all that the imagination built over and round it—we say too much for our own end and our own peace: and the end is not making, but marring.

She said unto Him, " They have no wine."

How reluctant even we are to have a rising intention forestalled! I remember the other day, Jessie, when you had ripened in your mind the scheme to take rooms in the Bankside cottage for poor Widow Burns, and I somewhat rashly suggested to you the very same idea, how you flushed up, and chided down some rising passion within, and one of life's great lessons, "Let alone," came like a wave over my heart's heart.

And so doubtless did the blessed mother repent of her words the moment after. For let it not be disguised that the reply she had from Him was in the tone of stern rebuke. It is not the mere address, "woman"—that He used in the last sad words on the cross—but it was the Τί ἐμοὶ καὶ σοί: What have I to do with thee? words which surely never would be used in other than rebuke. She had forestalled his first rising purpose by an ill-timed hint: and she was to be admonished that the high mission of the Son was not to be inaugurated at the bidding of the mother. But the purpose was not to be abandoned, nor the inauguration of the mission retarded. There is exquisite womanly tact in her words that follow. To the reproof, no answer: of the earnest purpose, no abatement. "She saith to the servants, Whatsoever He saith unto you, do it." Her instinct told her that He would do that, for the doing of which she was not to be his patroness.

And now, dear ones, I believe we have finished our Homily. It was not my intention to dwell on the wonderful miracle itself. Because it is so plain; especially to young and simple minds. "Let there be light, and there was light:" "let the water be drawn out wine, and it was so." It is one and the same power: that is all that is to be said: and in saying it there are volumes of blessing and comfort.

But notice one thing. The Lord Jesus acted here, with one of his beneficent bestowals, as He uniformly acts with the rest.

He created in abundance—lavishly—profusely. He created that which He has made for good, but which man's evil may turn into mischief. It is the dispensation of Eden over again : the tree in the midst, open and accessible. Thus God does, helping man with his grace. But how do *men* act in the same matter? Had Eden been man's garden, instead of God's, we should have had a cast-iron fence with spikes round the tree of knowledge : had some of our present philanthropists been guests at that wedding we should have had them beseeching the Lord of bounty and grace not to create wine that might inebriate, as we have them now trying to gain credence for a fiction that what He did create was not wine at all.

Well, darlings, let us be thankful in our bodies and our souls that God knows better, and that we are in his hands.

And so to Ken's Evening Hymn, to the old Tallis's Canon : the most soothing of hymns and of tunes.

VII

I CAN have no doubt who placed that little modest note on my bureau this morning. Guilty, little one ? No sign. Oh, I see, by looks exchanged between mamma and Jessie, that I was mistaken. Well, any how, the little note has gained the day, and shaped the fashion of our homily to-night.

But there is one thing in it with which I cannot comply. It requests that I will treat to-night the Lord's miracles of raising the dead. But if I take them all in one evening, I think we shall be not much the better for it. You know, there are three. Not that there were not many others : but three only are related for us in the Gospels. Of those three, the last, the raising of Lazarus, is immeasurably the most important : both in itself, and in the consequences to which it led. Besides, it happened quite at the end of the dear Lord's ministry here below : and I have several themes on which I should like to talk with you (yes, Margey, I feel the pressure of the little warm hand) before we come to the scenes at the end.

And as to the two others, I cannot consent to group them in one Homily. They present features so distinct, and so full of interest in their distinctness, that each must be considered alone.

We will take them in order as they happened. And notice that this is not the order in which they are found in our first Gospel, that of St. Matthew. I told you the other day, that there has been a great confusion of arrangement of events in the former part of St. Matthew. I shall have to say more on this point when we come to the second of our miracles, next time. At present, we will speak of the first.

The Lord had but lately finished that great series of discourses to the people, of which the Sermon on the Mount represents to us the substance. Turn to Luke vii. 1, and you will find the point in his ministry at which we are standing. There you will see that after finishing all those sayings, his first work was the wonderful healing of the Centurion's servant with a word : made more wonderful still, because the faith of the Centurion had anticipated this exercise of almighty power. Then, in verse 11, we find ourselves with Him on the following day. He has advanced onward on his circuit from Capernaum, and is approaching a place (*city,* in the Gospel ; but it was probably not a city in our modern meaning of the word) called Nain.

Now an idea of this approach may best be gained by my describing to you that to any hill village in the South of Europe : for travellers tell us that this was such a place, set on an hill.

Capernaum was on the shores of the blue lake of Galilee. From thence the Lord and his disciples had set out, possibly after the morning meal. They had toiled through the level and over uplands, along stony paths, sometimes quite rugged and unmade, sometimes, in the steeper parts, carefully laid with fitted stones, and curbs to carry off the water. To us, these stony paths are very wearisome: the continually varying surface twists the boot or shoe, and the sole becomes polished with the dry smooth stone, and slips back on the steeper ascents. But the bare foot suffers none of these annoyances. When once it has acquired a sole of sufficient hardness to resist the wear and tear of the journey, it accommodates itself to the hollows of the pavement, and it is never liable to slip. So that what would be to us a journey of some distraction from carefully picking the way, was probably

made by them without a thought of its difficulty, and with minds fully open to the
wondrous discourse of Him who talked with
them by the way, and made their hearts to
burn within them.

And so they passed on,—now in full blaze of
the sun on the side of the bare hill, now among
walled terraces shaded with olives and fig-
trees, now on the plains, full of verdure and
streams. On one of those terraces, there may
have been a mid-day rest and an hour of
slumber: or perhaps a discourse, or even a
miracle of healing, which has found no record
in our Gospels.

And now the sun was westering; and the
rocks over the olive-trees, and the tops of the
lines of wall supporting the planted terraces,
were lighted up with tints of orange and rose,
as they were climbing towards the eastward
facing gate of Nain. As they look up, a crowd
surrounds the entrance.

It is the solemn hour of carrying out the

dead: and there can be no doubt what that concourse means. On all their minds—yes, and on His among them, who was wrought upon, as we are, by common sights and sounds, falls a shadow of solemn thought, and a softened mood, shedding the dew of pity over heart and eye. The presence of sorrow—there is not a mightier power known. Of a thousand hidden places in the heart the secret doors at once fly open, and long-buried memories stir within. What a volume is human grief! Each of those apostles is in a moment busy with his heart's own bitterness, mellowed by the loving light of another's woe. Peter, the married man—John, loving and beloved—Thomas, the attached and desponding—nay, even one in whom two voices were as yet striving, if the Fiend had not ere this prevailed, even he may have felt the tear swelling up, as some memory of child, brother, parent thus carried forth, came rushing onward uncalled.

And He who walked in the midst—I believe

He did not on each of these occasions, any more than ourselves, necessarily know, or rather choose to know, recognise, and put before Him, all that He was about to do. He left Himself open to impressions, and let his resolves spring out of circumstances, on the surface of his marvellous being; though in the depths beneath there was perfect knowledge of what He would do, and of all things that were coming upon Him.

Who can tell how short a time it was since the head of the household at Nazareth, the just and God-fearing parent, had been borne forth on an evening like this? Who shall say what blessed thoughts then poured themselves over the heart of the son of the widowed mother,—thoughts of, and burning longings for, the day, when He who then held up her stricken form should proclaim Himself the Resurrection and the Life? And all these thoughts rose up anew on his soul at the sight of this funeral company.

And thus the two bands meet, and now the cause of the unusual throng appears. It is an occasion of no common mourning. The desolate is further desolated; a widowed mother has lost her only son. It is for this reason that much people of the city is there. They are doing what they can; poor service indeed, but still that which gilds the robe of sable with a hem of gold.

And now, dear ones, it is for us to fill up, if we would imagine that scene however insufficiently, the gaps in the Gospel narrative, by picturing to ourselves the action which prompted the utterances there recorded.

First comes the perception of the poor widow by the Lord. " When the Lord saw her." Now there is something worthy of notice in this name, " the Lord." It is probably never used of Jesus in this simple way by either St. Matthew or St. Mark. For Matt. xxviii. 6 is doubtful; and Mark xvi. 19 is part of an addition to the Gospel which probably was not written by the Evangelist himself. But St. Luke and St. John repeatedly use it, the former thirteen times, the latter ten times. It seems to have been a title which gradually came into use as the disciples learned more entirely to look up to and to pay obedience to Jesus. The practice of so calling Him had become usual before He said to them in John xiii. 13, " Ye call me ' Lord ' " (this is not in the vocative, Jessie, but is really " Ye call me ' the Lord ' "). And it forms a kind of title of dignity mingled with sympathy, the occurrence of which is sometimes very touching. So it is here. In this title " the Lord," here, there is a mixture of human sympathy and divine power. Who does not feel this when St. Luke says, " The Lord turned, and looked upon Peter " ? If " Jesus " only, the personal name, had been here used, how different would have been the effect—how shorn of its majesty —for " Jesus " in this scene is mainly the *sufferer :* if Christ,—how different—how shorn of its sympathy! But "the Lord" unites both these.

Well, "the Lord" saw her. Amidst the crowd, his eye, as the eyes of his disciples and of the much people of his own time, singled out this one central object, and the sympathy which had before been general, became special. The head of the one band who came for teaching and witnessing and beholding, approaches the head of the other band, who came for mourning and comforting.

But what jarring words are these, "Weep not!" How often are they spoken helplessly and in vain! That can hardly have been so in this case. We must suppose that there was something in the Lord's look and gesture as He spoke them, which carried strange feelings into the poor mourner's heart. There must have been a majestic drawing up of the form— a power kindled in the eye, of Him who spoke these commonplace words of conventional comfort. They were less a consolation, than a command; which to obey, was consolation. And that witness of the fact who was induced to call Jesus " the Lord " in describing it, has by this very name set before us this influx of his divine power on the gushing of his human sympathy. "The Lord" is a living token that he who first wrote the narrative, had seen the transaction with his own eyes and was one with the spirit of it.

And now, darlings, let us follow on. The sad mother, we may well believe, stands still in her astonishment, waiting for what is to come of this word of power, which has staunched her grief within her. Nor does she wait long. All eyes are upon Him who has thus interfered with the sacred flow of sympathy and woe. Some may have regarded Him merely as a rude disturber : but I doubt whether these were many. Some may have had faith enough to anticipate what He was going to do: but I am sure these were few. That Centurion who is even now thanking

God in his joyful household—that blessed
Mother who went beyond herself in prompting
his first miracle—I know not whether there
were any others in whom there was such
faith : and neither of these was present. ˙ The
greater part of those who were, were probably
in a doubtful state of mind, not knowing how
far his power might extend,—expecting something, they could hardly shape what.

But doubt is soon clearing up and giving
place to eager expectation. His next approach,
his next address, is not to the mourner.
The reason why she is not to weep is not to be
given in words. " He came and touched the
bier."

The body was being borne out, as we may
even now see the dead carried forth in the
South of Europe, open, in a bier, the face and
the hands uncovered. As yet the procession
was moving on. But there was that in the
mien of Him who had his hand on the bier
which at once stopped the bearers. It spoke,

of high resolve, and power to accomplish it. It was a look, in the presence of which no ordinary matter could go on. How many hearts are beating high in those two multitudes, now mingled into one! Upon how many minds is breaking the truth that One is present who can call the dead to life! Of all perhaps she who is most concerned realises this the least. Immense issues like this rather numb the spirit, than excite it. She stands, awed, calmed, but scarce consciously expectant.

But what words are these? "Young man, to thee I say it," . . . . Person addresses person. It is no new creation, observe. The Almighty One is not about to call a new life into being. The poor body, already touched with decay, is still, in the eye of God, the tabernacle of the soul:—"Young man,"—or is this an anticipation on the part of the Lord of that re-union of body and soul which is so soon to take place? "To thee I say it"—

to thee, wherever tarrying in this disembodied state, but by the word called back to the body, " ARISE ! " Enter again, and take possession : be fitted, each minutest root and fibre of organized nervous life, into that junction of which decay has already blunted the fine points of union—let the muscle be knit into its elastic spring, and the brain, collapsed and dethroned, re-assume its central sway. All this, and a thousand other hidden wonders, must take place, before the dead man can sit up as the Lord commands.

Yet all *is* done, in a moment. He who made can remake. Yea, more—life returns, not by instalments, but at once and complete. In the beautiful story of Alcestis, which I read to you, Jessie, the other day,* you remember, that when Admetus has recovered his lost one from the grave, she remains silent until the third day :—speech is the last sign of

* From " Chapters on the Poets of Ancient Greece ;" a book he published, 1841.—ED.

renewed life which returns. But here it is not so. "The dead man sat up and began to speak." We are not told of what. Did he question his recovered being? Did he wonder at the throng, the amazement in all around him, the strange evening light? Did he stretch his arms to his mother with a cry of raptured joy? We cannot tell. "And He delivered him to his mother." "See, thy son liveth." Oh what joy ineffable! What bright spots must there have been in the course of the Man of Sorrows, when He was able to cause, and to witness, so much joy!

"There came a fear on all." What had they seen? in whose presence had they been? A great Prophet had indeed risen up among them—greater than Elijah and Elisha, who stretched themselves on the dead children and prayed: greater than any of whom Scripture had hitherto told. And this they further expressed when they said that God had visited his people: for by those words did Israel commonly speak of the coming of the expected Messiah.

One word more. It was this miracle among others to which our Lord appealed, as we learn from what follows in St. Luke, for proof that He was He that should come, and men were not to look for another.

Oh wonderful power and wonderful love! just one little sample of what the dear glorified Lord can do, and will do! Love Him, my darlings, keep ever close to Him in thought and affection. And then no death shall ever part us. He will come in the world's evening, and stand on our earth, and lay his hand on the grave, and say, " I say unto thee, Arise! " And then we shall sit up and begin to speak his praise: and He will deliver friend to friend, and child to parent, and dear lost ones to their widowed mates—and all earth will shout, and all heaven will echo, that GOD HATH VISITED HIS PEOPLE!

VIII

WE are to speak this evening of the raising of Jaeirus's daughter. Of the three miracles of raising the dead, that at the gate of Nain is the most wonderful—the bringing Lazarus out of his tomb the most awful: but this the most deeply interesting.

The Lord had just come from his visit to the Land of the Gergesenes, where he had cast out the devil from the man, or men, who dwelt in the tombs. This is plain from the very careful and precise notice in St. Mark and St. Luke. In St. Matthew, all this part of the history is in confusion. He makes the message of Jaeirus to Christ take place while He was speaking to the people, after the question about fasting asked by John's disciples, which in reality happened a considerable time before, as you will see by comparing the carefully arranged narratives of St. Mark and St. Luke.

When Jesus returned the multitude gladly received Him. They were waiting for Him. The last they had heard of his teaching was that magnificent series of parables concerning the kingdom of God which we have most fully detailed in Matt. xiii. For, however little we might suspect it from St. Matthew's account, it was on the evening of that very day, as specified by St. Mark, that He crossed the lake to the country of the Gergesenes.

The people were anxious to hear from Him more of the same kind. For there is this in the parable, which was now his way of teaching,—that it speaks to every hearer as he or she can receive it. The little child enjoys the story, and learns something. The plain man enters into the worldly wisdom of it, and learns more : and the wise and more thoughtful the listener, the more would the wisdom of the things said reveal themselves. No wonder then, that the people flocked to Him, and pressed upon Him, to hear the word of God. So they anxiously expected Him, all on the shore waiting his arrival.

But there was one who had been waiting with more anxiety than the rest. Jaeirus, the ruler of the synagogue at Capernaum, had an only daughter, twelve years of age, lying at the point of death. All human means had failed, and there was but one who could save him his child: and He was absent. But on the Lord's arrival he at once hastens to Him.

And here notice again that St. Matthew represents the ruler as coming in, and worshipping our Lord, and saying that his daughter was dead: whereas we know from the other accounts that she was not dead, but dying. And this of course makes an immense difference in the nature of the man's faith: for in St. Matthew's account he says, in spite of her being dead, "But come and lay thy hand on her and she shall live:" whereas, in the longer and more exact accounts, it was our Lord Himself who suggested to the poor father that even her death was no reason why he should be afraid, but only believe and she should be made whole.

Now some might ask, and I know mamma herself sometimes inquires, why I bring forward these differences between the Evangelists —why I do not rather conceal them, or, if I mention them at all, adopt some ingenious way of making out that they mean the same, though their words are different?

Simply, my darlings, because I believe that I should be dealing unfaithfully by God and by Truth in doing so. If He has been pleased that the Evangelists should give us differing accounts of the same fact, it was for wise reasons that He did so, and that we might make wise use of the difference, not that we might cover it up and hide it out of sight.

To my mind these differences (discrepancies, as they are called) are the strongest possible marks of the truth of the facts themselves. If the Evangelists were deceivers, intent upon making men believe that things happened which never did happen, they would have taken good care that no such differences should be found in their stories. But being, as they were, independent and honest narrators of facts which really happened, they were liable to what occurs to all human witnesses—they reported variously, and sometimes inconsistently one with another. And there used to be no difficulty in getting people to acknowledge this. I remember in the Greek Testament which we used to read when I was young, with notes written by a very good sound man, Dr. Burton, Professor of Divinity at Oxford, we were told regarding some difficulty in the Acts of the Apostles, that " probably St. Luke was not well informed respecting this part of St. Paul's life : " and no one thought any harm of this then, because it was—not perhaps in this particular case, but in many others— what every intelligent man then thought. But men are now become such slaves to the letter, and so careless about the spirit, of Holy Scripture, that they will maintain every account to be separately true in every detail, or else they say we cannot believe our Bibles.

Well, darlings, we all do believe our Bibles : and yet we do not believe that Jaeirus's daughter was dead, as St. Matthew says she was, when her father came to Jesus.

Now Jaeirus was the ruler of the synagogue, and as such was perhaps present when the Lord healed the withered hand some time before. His faith had been deeply rooted before he could say of his dying child, " Come, lay thy hand on her, and she shall live." You know in some measure, dear children, and mamma and I know in full bitterness, the anguish of despair which takes possession of us when one very dearly loved lies in extreme danger. And *we* have felt it all the more, because it has so happened that not one of our beloved ones has ever been seriously ill and recovered. In their cases illness has meant death: and so, as one after another has been struck down, we have sunk at once, I fear, into hopelessness from the first—hopelessness which it would take a very strong faith to penetrate and illumine.

The blessed Lord was never slow at the call of faith. He, followed by his disciples, went at once with the anxious father. But on the way an incident occurred which occasioned delay. And the delay, humanly speaking, was fatal. During it the poor child had died. " It is of no further use to trouble the Master to come;" thus ran the message from the ruler's house. What effect the words had on him we are not told. Jesus overheard them being spoken, and made answer to them by an address to the father, " Be not afraid: believe only, and she shall be made whole."

The inhabitants of Capernaum could not have been ignorant of the raising of the widow's son at Nain : so that we can hardly attribute the tenor of the message to Jaeirus to ignorance of the *power* of Jesus. We may fairly set it down either to the slowness of men to take in and appropriate (so to speak) such astounding facts, or to some idea that that former miracle had been wrought in favour of a peculiarly heartrending case of bereavement, and that it formed no justification for urging on "the Master" a request to do the like in this more ordinary case.

But the company have reached the house : and there they find the noisy accompaniments of Eastern woe, "the minstrels and the multitude making a noise," as St. Matthew has it : "a tumult, and the people weeping and wailing greatly," as St. Mark. It does not quite appear whether all the disciples, or only Peter, James, and John, came with Him : probably only the latter, as apparently asserted by St.

Mark. But however few the attendants might be, the intrusion of strangers at such a time, with a purpose manifestly alien from the mournful employ of those assembled, set the two companies in strong contrast. Their behaviour provokes a rebuke from the Lord: " Why make ye this ado, and weep? the child is not dead, but sleepeth." All the Evangelists describe in the same words the reply of the multitude: " they laughed Him to scorn;" and St. Luke, the physician, adds, " knowing that she was dead."

But He put them all aside. There was something of the majesty with which He had gone up to the bier at Nain, when they who bare it stood still: so that all yielded and made way.

But what exactly do we suppose the Lord meant by " She is not dead, but sleepeth?" Because, you see, she was really dead: and the knowledge that she was, provoked that scorn of which we read in the minds of the people. The meaning seems clear, if we think of what He had it in his mind to do. " This death, which you weep and bewail, is in reality no more than a mid-day slumber, so soon shall it come to an end."

And now they enter the chamber of death, those six—the Lord, the three disciples, the father and mother of the maiden. How like are all chambers of death—and yet how unlike others was this one! You remember when we four last entered such a chamber *—and on that little press-bed in the corner by the window lay all we cared for in that room. The rosy beams of the setting sun, I remember, streamed in at the window and filled it with glory; the sea outside lay bright under the great red orb : what recked we of these, or of aught else there but of that one pale form ? We scarce dared breathe — even grief was lulled, and all was solemnised, without a feel-

* This description, like the one in the Second Homily, refers to the scene of our dear boy's death in 1850.—Ed.

ing beyond. And, as I said, how different was this chamber of which we now speak! There lay the pale form from which life had but now fled, with, we may well imagine, that first sweet look of death, which seems more like the living person than was even life itself:—

> " Before decay's effacing fingers
> Have swept the lines where beauty lingers." *

But here is something more than that pale form : here is One greater than death : here is a restless hope, a leaping expectation, that makes every heart throb, and fixes every eye, not on the lifeless form, but on Him, the Lord of Life. See—He approaches the bed—He takes the waxen hand as it lies on the side— He speaks, and they are the gentlest word of affection. " Come, my child," He says— no more; and the maiden arises as if out of a sweet slumber at awaking time. She rises,

* From Byron's " Giaour," line 72.—Ed.

not to languor and disease, but to health and appetite. She begins to walk; and He commanded that food should be given her.

And all this is to be kept secret. " He ordered them many times that none should know the matter," says St. Mark. Very strange sound these injunctions of the Lord. For how was the truth to be kept from the multitude without? What account could the parents give, but one? The death might no doubt have seemed to them to have been but a suspension of life: but here was more. Here was the girl who but now was in the last stage of dangerous disease, in perfect health. She who had not perhaps left her bed for many days, was now walking among them, and strong. It may seem strange to say it, but one cannot help feeling that these injunctions of the Lord must always have been from the very nature of the case, broken—as we know they were broken in some related instances : " the more He commanded them, the more a

great deal they published it." And the real
practical result which they had may have been
this—that they lay as an obligation on the
consciences of the healed person and the
friends, not to make vain boast of the miracle,
or to use it as food for idle curiosity.

IX

THIS is an evening I have long looked forward to. For of all the narratives of our Lord's acts none is so full of his majesty and his sympathy as this of the raising of Lazarus. We have been in some measure prepared for it by those previous miracles of raising from the dead; but when we approach this, we at once feel that we are in a different atmosphere, so to speak, and with wholly different circumstances to consider.

Let me try to explain this. Those other two were wrought, so to speak, in the course of events. The Lord was sent for to the house of Jaeirus: He met the funeral procession at the gates of Nain. Doubtless both wonderful acts did much to disperse his fame abroad.

Of the latter of the two we are expressly told this.

But with this raising of Lazarus all was different. It was not a mere incident in one of our Lord's progresses. It was a matter carefully prepared and brought about by his own design. It was, again, not merely an act which spread abroad his fame. It had indeed an effect, but of a far other kind. It was, humanly speaking, the great moving cause of the Lord's death. And He deliberately did it, knowing it would have this effect.

Now all this removes it away from his other miracles in dignity and significance. And in another particular also is it removed from them. You will remember, that not a word is said of it in the three first Gospels. We have it from St. John alone. And this is a circumstance full of strangeness: so much so, that many have been thereby induced to seek for some reason why it should thus have been. But no reason has ever been found,

worth my naming to you : none which would
for a moment satisfy young minds accustomed
to require something like probability when
reasons are given.   Whatever may be the
cause, so it is : that neither in St. Matthew,
nor in St. Mark, nor in St. Luke, is any hint
whatever given of this greatest miraculous act
of Jesus, of this which, above every other act,
brought about his death.   St. Luke admits us
to a glimpse of the family, and gives us an
interesting trait of the characters of the two
sisters ; but he never mentions their residence
at Bethany, nor indeed that Jesus went out to
Bethany during the former nights of the week
of his passion : which is related by the two
other Evangelists.   Nor does St. Luke mention
the anointing which took place in the house
of the sisters : an incident related by the other
three.   These facts may serve to show us how
vain is the attempt to give any account of the
framing of the Evangelic narratives.   Thus
much indeed we may say, that it was natural

for that Evangelist who gives the inner history
of our Lord's life and acts, to relate in all
its detail a matter so full of the manifestation both of his humanity and his divinity;
natural for one who traces onwards his glori-
fication through the conflict which ended in
his death, to dwell on the great act which
brought that conflict to its issue.

What, little one, has the head sunk on my
knee already? The beginning is not promis-
ing, is it? Too dull, and too many long
words. Well, we'll try now to come to something simpler.

The Lord had withdrawn himself from the
enmity of the Jews, which had risen to a head
after the healing of the man blind from his
birth, and the discourse about good and bad
shepherds. He was on the other side of
Jordan, probably at the other Bethany or
Bethabara, at the distance of about a day's
journey.

And now let us look in on the family at

this Bethany. Here are a brother and two
sisters, all beloved by Jesus. They are not
called his disciples; but they evidently believed on Him. They seem to have been
people of some wealth. We find one of the
sisters anointing the Lord with very costly
ointment: we find chief men among the Jews
coming to visit and comfort them. All this
is strangely interesting to us. Were there
more such persons? or were these the only
persons thus beloved by our Lord?

There was another member of the household
of whom we hear nothing in this history: and
that member was its head. For it was " in the
house of Simon the leper," as we know from
St. Matthew and St. Mark, that the Lord was
anointed by Mary. Was Simon identical with
Lazarus? Or was he no longer living, and
was the house still called by his name? Or
was his house, when the anointing took place,
not that of Martha and Mary and Lazarus?
and were they found in it merely as neigh-

bours or friends? Or, again, was he perhaps the husband of Martha? This may well have been: for Lazarus evidently was not the master of the house, being only one of those who sat at meat with our Lord on the occasion of the anointing.

So there is, you see, some uncertainty about the members of the little household: which however does not affect our conception of the three concerned in this history — Martha, Mary, and Lazarus. Now we all know the course of the sacred story. I am not going to tell it over again, but merely to try to give life and freshness to our conceptions of it, by dwelling on some portions.

Remember what we read last night in preparation for our present employment — that beautiful and living description in Stanley,* of the situation of Bethany, and the roads leading from it in each direction. And, remembering this, let us stand with the two

* "Sinai and Palestine," pp. 189—195.

sisters at the sick-bed of him whom Jesus loved. In vain they look,—in vain they have for one whole day looked,—along the path which ascends from the deep valley of the Jordan for the well-known form of Him whose coming is to end their harassing anxiety. Somewhere under that distant line of mountains He is tarrying. How should He have received their earnest call, and not have obeyed it? Of his power they have no doubt. If He were here, their brother would be safe. This heaving frame would be quiet, these damp dews of death would disperse, at the entrance and the first word of Him whom they had long believed to be the Christ, the Son of God, who should come into the world. But He comes not. If there is no distrust of his power, there gradually arises something like distrust of his loving care. The words of the two sisters to Him when He does come seem to show this—" Lord, if thou hadst been here, my brother had not died." But the dreary hours of hope deferred pass away, the dark moment of death itself is past,— the burial in the cave outside the village—the first day of bereavement,—the second,—the third,—with their companies of sympathizing and well-meaning comforters from Jerusalem coming and going: and yet—no help!

At a certain time on the fourth day, Martha, busy about her household matters, or having gone forth into the village, hears that the figure of Jesus has been descried coming up the path. In a moment, without waiting to tell her sister, who was in the house with the band of friends from Jerusalem, she runs to meet Him. And now, I think, there is something to be said about Martha's words to Him. First comes the sad and somewhat complaining sentence: His presence might have prevented that which had happened. Of this it may of course be said that it *need not* convey a complaint—because, if the death took place before our Lord could possibly have reached Bethany (which it is not unlikely it did) the words may merely be an expression of earnest faith in his power, whenever and wherever present.

But what does Martha add? "Even now, whatsoever thou shalt ask of God, God will give thee." What does this mean? I won't confuse my little listener with other meanings than that which I must think is the only right one. Martha, in her forwardness and rashness, for an instant anticipates the great work which the Lord was about to do—and she means, "God will enable thee to raise him up again, even now that he has been dead four days." It was truly a rash thought —and one not soberly formed by herself. Afterwards she loses sight of it, as hasty spirits are apt to lose sight of their own shifty moods : she is blind to it, nay, she even recoils from it, when it was on the point to be realised.

Have we no similar instance in the Gospel

record? Was not the rashness of the Lord's own mother at Cana in Galilee something very like this? And do not our own lives furnish, in their lower degree, something of the same kind? Whilst some hardly certain step is hanging in doubt in our own minds, and we are balancing interests and probabilities, is it not frequently found that a woman or a child has anticipated, by a swift instinctive leap, the final result of our deliberations, and announced our intention long before we had formed it? Not that the Lord was thus wavering—He knew what He would do: but the analogy is good on *one side* at least, that of her whose rapid anticipation in its very weakness grasped the great conclusion, pre-determined by the Life-giver in his strength.

Understand her words thus, and all follows naturally. "Thy brother shall rise again." "Yes," replies Martha, somewhat baffled and thrown out of her train of thought by the words, "I know that he will, when all shall rise at the last day." And how is this natural? Why does the Lord thus put her off? Simply for this reason. Her previous words placed the matter on a wrong ground. They pointed at the bringing to life again of her brother. True: but merely as a favour to be asked of God, and to be obtained, as a very holy man might obtain it, by prayer. Such was not the thought with which the Lord's great act was to be approached. His own power, granted Him by the Father, but not as a mere answer to a human prayer may be granted, to have life in Himself,— to be the Resurrection and the Life, the annihilator of Death in the body and in the soul,—this must be put forward and acknowledged, before the glory of God in this act could be seen aright. Does she believe this?

Well, she does not quite apprehend the glorious words: she is asked one thing, she answers another thing: she is ready to believe all—for she has long believed that He who speaks to her is the Christ, the Son of God who was to come into the world. Whatever this implies, she believes; but neither in her answer, nor in her thoughts, does she enter into the full meaning of the Lord's question. And now she goes at once and calls her sister. Perhaps the Lord had *at this moment* asked for Mary: perhaps He had done so before, and Martha naturally, when He began to treat of high and holy truths, thought of fetching her of whom we know from St. Luke that she was used to sit at his feet and hear his word.

She called her *secretly.* That is, I suppose she sent in to the place where she sat with the Jews, to tell her that she was wanted: and then, on her coming out, imparted to her the intelligence. That this was probably so, is shown by the Jews only remarking that she rose up quickly and went out.

They found Jesus in the same place where Martha had met Him: a plain proof that it

was really He who had sent for Mary, as
He remained there awaiting her coming. We
can all understand why He would not enter
the village,—without supposing, as some have
done, that it was because He desired privacy.
This was in fact the only one of his miracles
for which He desired publicity, as we shall see
further on.

The place was in the way from the village
to the grave, as appears by the Jews, who
thought Mary was going there to weep, following in her track. As she went, she wept,
and those who followed her wept also. And
here we have the very deepest and holiest
proof of the Blessed One's full share of our
common nature. The sight of sorrow in
others, if that sorrow also be his own, is more
than even the strong man can bear. And
the Lord's act of self-command is described
by a remarkable term: the same word as is
employed when He rebuked those whom He
healed, and charged them not to make Him

known. If such an English combination were allowable, " vehemently checking himself " would seem to be nearest to the sense : at all events, nearer than " groaning," as our version has it, which is not at all what is intended. This check seemed to be necessary for the utterance of the question which followed : " Where have ye laid him ? " But no sooner is it uttered, than nature had her way. By those tears, human sorrows are hallowed, human bereavements are glorified. The gods of the heathen might not look upon death : the world's Saviour wept at the grave of his friend. Which of these would draw all men unto him ?

On the great scene of the miracle itself, I say but these few words. Martha seems utterly to have shrunk back from her momentary hope, when she deprecates the opening of the tomb. Her words may even have been intended as an express repudiation of her former rashly uttered assumption.

K

Secondly, the Lord's words must be regarded as having a close connection with that former utterance of Martha's. He knew that the Father always loved Him; and in this case too his prayer had been answered. So that his thanksgiving was not as for anything unexpected or unusual; but had for its object to persuade those who stood around of his divine mission.

Thirdly, that the Lord's call to Lazarus was not merely " a loud voice," but a loud shout—symbolizing and anticipating that voice of the Son of God which shall echo one day through the sepulchres of the world.

As to the manner of the dead coming forth, some think that the grave-clothes were only swathed round each limb, leaving movement free. But it seems far more probable from our Lord's command, " Loose him and let him go," that there was, as an old writer observed, a miracle within a miracle—that the form, swathed and confined, glided forth supernaturally from the tomb. And so the most ancient pictures represent it.

I have observed before that the Lord, who in the earlier days of his ministry shrunk from publicity for his wonderful acts, now courts it. He even utters his prayer " because of them that stood by." And all the consequences of the work, wrought under the eye of unsympathizing enemies, wrought at the approach of the great assemblage of the Jewish race at Jerusalem, were before Him : the raising thereby of the enmity of the Chief Priests and the Pharisees to its highest : the fatal decree of the council which followed on the miracle : the bringing about of the tri- umphal entry, which St. John attributes en- tirely to this miracle : the week of conflict, ending in the cross and the grave.

And we seem to see that all this had been long present to the Lord's thoughts. For in the parable of the rich man and the beggar, we have remarkable anticipations of it. We are told there of hardened ones who had in their hand Moses and the prophets, but believed them not: who would not be persuaded, though one went unto them from the dead. And, strange to say, in this one alone of the Lord's parables, is a name given to the dead subject of the history, and that name is Lazarus.

And now, darlings, this has been, after all, a disappointing evening. Of all the Lord's miracles, this one, I suppose as so far transcending in its sympathy and its majesty all our thoughts and powers, most completely baffles the preacher and writer. I never heard a sermon, I never read a comment, on the raising of Lazarus, that did not send me away empty, as compared with the fulness and refreshment after the Evangelist's own grand and simple narrative. And so let our night's lesson be, how empty are our thoughts, our words, our feelings, in proportion as his

Presence is manifested, who alone has all fulness in Himself.

Jessie will perhaps repeat the touching lines of our Laureate,* which occur to us as we leave the wonderful story.

> When Lazarus left his charnel-cave,
>   And home to Mary's house return'd,
>   Was this demanded—if he yearn'd
> To hear her weeping by his grave?
>
> " Where wert thou, brother, those four days?"
>   There lives no record of reply,
>   Which telling what it is to die
> Had surely added praise to praise.
>
> From every house the neighbours met,
>   The streets were filled with joyful sound,
>   A solemn gladness even crown'd
> The purple brows of Olivet.
>
> Behold a man raised up by Christ!
>   The rest remaineth unreveal'd;
>   He told it not; or something seal'd
> The lips of that Evangelist.*

---

* Tennyson's "In Memoriam," No. 31.—ED.

X

WHAT shall it be to-night? Well, Margey, there are many things tempting us— which shall we choose? Our last talk was about that wonderful raising of Lazarus. But we advanced thus far in our Lord's life only in order to put together those three raisings of the dead which are recorded for us in the Gospels. We have omitted much that might serve to bring his blessed words and works before us on these our Sunday evenings. Shall it be miracles, or parables, or incidents? We have a great work before we finish if we are to talk of the Lord's sufferings and death and resurrection; but we must necessarily reserve these till last. Let us pause and think.

Well, it seems we ought to have thought

before. So suppose we give ourselves a week, and by next Sunday get our plan arranged, and meantime I will take a subject closely connected with our great theme, the Lord's life and words, though not in its direct course; I mean the course of him who was the Lord's forerunner, John the Baptist.

We have already said something about the beginning of that course when we spoke of the childhood of Jesus. The two holy children were relations : what relations, does not appear; for the word rendered " cousin " in Luke i. 36, is merely " kinswoman " in the original. They might doubtless thus have been frequently thrown together. This has perhaps been assumed in our thoughts more as a matter of course than it should have been. In childhood, and as long as Elizabeth his mother lived, it may have been so ; but she was old at John's birth, and so was his father, Zechariah, and the words, " the child . . . was in the deserts till the day of his showing unto Israel," would seem best interpreted by supposing that when he became an orphan he took to a wild solitary life. At all events, at the time of the baptism there does not seem to have been any very intimate acquaintance between him and our Lord, as we shall have occasion to remark by-and-by. Still, there appears no reason why there should not have been frequent intercourse in childhood. Let us think on this first.

Of course the representations of Christian art are symbolic, rather than aiming at matter of fact. As we before remarked, the Holy Child is ever the unclothed and sinless One, whereas John is partially clothed to represent the fact that he had inherited sin, and therefore shame. Probably in matter of fact, both children were naked alike without clothing; for such is the custom of those Eastern countries, and their customs do not vary. Nor, of course, have we any right to assume that there was any such constant pointing out of the Child Jesus as our painters are fond of representing; nay, we may be certain there was not: otherwise John could not in any sense have said, "I knew Him not" at the baptism. This again, in our paintings, is purely symbolic.

I make these remarks to prevent our confounding truth in art with truth in fact. That may be intensely true in art which never happened, and could never have happened, in fact.

Well, the aged father and mother are gathered to their rest, and the strange boy is left alone. He had "waxed strong in spirit;" was bold, and solitary, and original; unlike other children. We may conjecture by the after-events of his course, what was the character of this originality and strength of spirit. The conventional sins of his time, the pomps and vanities of society, the corrupt degeneracy of the people of God—these were to him matters of aversion and loathing. Instead of soft clothing worn in kings' houses,

he is content with the scanty girdle of skin,
and the cloak of camel's hair for cold; instead
of delicate fare, he feeds on the edible locust,
and flavours it with the honey from the rock.
His dwelling is in the den, or cave, or tomb;
on the mountain side he roams and meditates.
Rumours are abroad that he has been seen by
one and another; we may imagine him, before
"the day of his showing unto Israel," going
far to warn the sinner hastening to his ruin,
testifying to right, telling the truth, and
boldly rebuking vice; we can fancy his fame
spread far and wide as a strange and holy
youth and man (for near upon thirty years
thus passed), before the Spirit of the Lord
came upon him. Thirty years — or say,
twenty. Do we realise this? Imagine that
a lad in strange dress and uttering strange
words, had taken up his abode in our forest
in 1849, and was living there as a man still.
We can hardly imagine the state of society in
which such a thing could be.

Well, the years sped on—and at last the young man's thoughts, so true, so noble, so wild and solitary, were lit up by fire from above. The word of the Lord—the "daughter of the Voice," as the Jews called it, came to him in the wilderness. And then he saw that he was set not merely to rebuke the age and to warn sinners, but to prepare the way of the Lord; that the old prophecies which he had learned at his holy mother's knees were to be fulfilled in himself and in One whose herald he was to be.

And then the wild figure came forth from its hiding-places, and began to cry, " Repent, for the kingdom of heaven is at hand." Up and down, in the tracks of the passover pilgrims, along the march of soldiers, beside the caravans of merchants, the weird Elijah-like form appeared—the loud piercing cry sounded. We can see him, as old painters have given him, rough and desert-grown, but beautiful in the strength of his youth, bare to the loins,

and with his sinewy arm pointing to heaven.
And some doubtless scoffed, and others went
their way quick for fear they should look and
have to listen ; but more stood and heeded
him.   And then, what a scene was there !
For not one did he spare. No compliments—
no soft words.   The merchant, the soldier, the
publican, all heard the bold words that went
right home and rankled in the heart.   And
when he saw some of rank and learning
coming, it was not, "Make way for those
gentlemen ;" but it was, "O brood of vipers,
who hath taught you to flee from the wrath
which is at hand ?"

But what do we see ? The holy man enters
the river Jordan, which flowed by the scene of
his crying in the wilderness ; and hundreds
enter with him.   One after another they cast
off their garments, they bow the head, and
pass under the water at his bidding—the
water which represents to them their pro-
fession, repentance and the washing away of

sin. But as he baptizes, he protests—"This is not all—this is not the end—this baptism is only a sign of repentance—there is another and a greater—a pouring out of fire and the Spirit of God—and One is coming to perform it, One who is more above me, than the master is above the meanest of his slaves."

Strange scene—to which no revival on this or the other side of the Atlantic has ever seen the like—Jordan's banks and Jordan itself full of eager multitudes—all the Holy Land there, and crowds from outside it. What an era it must have been in the "religious world" of that land! How old conservative Scribes and Pharisees must have shaken their heads, and called out upon the times! How many decorums must have been violated, when wise old grey-beards submitted to be taught by this wild rustic, and dainty daughters of Israel passed through the waters under his hands weeping tears of penitence!

But such, dear ones, is God's way when He

prepares it; and doubtless many a score of disciples was born for Christ who came after, by this revolutionary infringer of decorums. For his cry ever was, "Not I, but another—a greater than I is coming! My baptism is but half—is not half, that which God has to do with you."

One day, among the crowd, came a young man to be baptized, known to John, and yet unknown. Known—for he had long been acquainted with Him personally—he had long known his blameless life and holy character; but unknown, for he had never heard of his high mission; this was known probably, since the death of Joseph and the parents of John, to none but her who had laid up and pondered all the events of his infancy in her heart.

With noble humility, John declares, "I have need to be baptized of thee: and comest thou to me?" And after the baptism, a sacred sign, long ago announced to him, reveals to

him that his holy kinsman was the Greater
One, who should baptize with the Holy Ghost.

At once the self-renouncing Baptist pro-
ceeds to fulfil his direct mission, of sending
men to Christ. No long time after, when a
commission was sent out from Jerusalem to
demand of John who he was, and why he did
these things, he plainly told them that the
Great Baptizer with the Holy Ghost was
among them; that he was nothing but a voice
in the wilderness, preparing the way before
Him. And the very day after this inquiry,
he directs two of his disciples away from himself to Jesus, pointing Him out as the Lamb of
God.

There is nothing finer in history than this
self-renouncing : except perhaps the noble
testimony which the same holy man bears to
our Lord further on in the same fourth Gospel,
calling Him the Bridegroom, and himself only
the Bridegroom's friend — announcing that
Jesus must increase, and he must decrease.

And accordingly from this time onwards the star of the Baptist wanes and pales before the rising Sun.

The next we hear of him is in the court of a licentious tyrant, Herod Antipas. This bad man, as it so often happens, was not wholly bad. He had times of good resolves, and at such times he sent for John, and heard him willingly, and even made changes in his conduct in consequence. But at last, as so often in intercourse between bad men and good, came a time when something had to be told which would give mortal offence. Herod had taken away Herodias, his brother's wife, and lived with her as his own. "It is not lawful for thee to have her," said the intrepid man of God : thereby making himself the object, and at last the victim, of the abandoned woman's vengeance.

But first comes a very remarkable passage in John's course, respecting which some have doubted as to how we are to explain it.

He was lying in Herod's prison at Machærus, a town in Peræa. Our Lord was teaching and working miracles in the neighbouring district of Galilee. Let us think for a moment of the two.

We may well imagine what an earnest and lively interest John must have taken in the Person and career of our Lord. Think what his own baptism had been—and then remember that it was to be as nothing in comparison with that nobler baptism with the Holy Ghost and with fire which the Lord was to confer. His own, the lesser baptism, had been frequented by immense multitudes from the whole of Palestine. What then was this second and greater baptism accomplishing? In his prison he heard of the works of the Christ. But what were they? A few miracles of healing in an obscure district—and these accompanied with strict injunctions to the healed not to publish his fame abroad. Was this He who was to come? Not that the Baptist could really

doubt concerning the Person or the mission
of Jesus. But all this was so unlike anything
which he had foretold—the Lord's way of
comparative obscurity and delay was so incom-
prehensible to him—his soul too, very likely,
was so faint, and sick with hope deferred,—that
he sends by his disciples a message, meant, I
cannot doubt, as a stimulus to that glorious
course which seemed to him to be but half
undertaken, "Art thou He that should come,
or do we look for another?" How the Lord
answered this inquiry we know: our present
concern is more with the testimony which it
drew from Him to the person and career of
John. This reed shaken with the wind, this
Fainter under persecution,—was John this,
when they went out into the wilderness to see
him? Nay, he was a prophet, and more than
a prophet: the greatest of the sons of men,—
the messenger sent to prepare the way of
the Lord. But the grace of the Gospel, the
standing and glory of even the little ones of

Christ's kingdom, these he had not : the brightest star of night is weaker than the feeblest beam of the risen day.

And now we come to the closing scene—that dark record of hasty vows, and watchful vengeance, and holy blood shed as water on the earth.

It was a great festal day. The lords and high captains and chief men of Galilee sat at a banquet with the king. The feasting was over, and the wine was flowing freely. The daughter of Herodias came in and danced before them. Strange place and scene—and strange consequences followed. The infatuated king, robbed of his better senses by the pleasure of the licentious exhibition, made her the fatal promise,—confirmed it by the fatal oath. The girl is puzzled what to ask—she goes and consults her mother. Then vindictive hatred, ever ready, makes its spring : " Ask for the head of John the Baptist." And so, as a bit of after-supper entertainment, at the asking of a worthless girl, in spite of the wretched king's undissembled grief and reluctance, the minister of death is sent to the prison, the greatest of the sons of men is foully slain, and the holy lips which had uttered the message of God's Spirit to Israel are given into the hand of the wanton, are passed about for the gaze of the revellers, and are finally delivered up to the criminal wife herself to insult and cast out.

Never was a more terrible tale—never one with sadder, wilder contrasts. It has almost the same character, in its higher realm of interest, as that of some of the old Greek histories of which I was telling Jessie the other day—the character of irony—mockery of human pomp, and mockery too of even solemn human hope. The sport of the reveller, the reward of wanton frolic, is the mangled head of God's prophet: and on the other side, the end of the glorious career of the forerunner of the world's King is to fall as it were by a chance blow, victim to the spite of a worthless woman.

But if the holy martyr had small respect there, it was not so everywhere. The poor mutilated body was cared for by the disciples who had followed his teaching : and when they had done this, they went and told Jesus.

And here is perhaps the most notable feature of the whole. The Lord when He heard it, went apart into a desert place to pray. Was it, that the Forerunner had gone before Him, as in ministry, so now in suffering? For from this time seems to come gradually over Him that deepening cloud into which He entered further and further, till He fell into the very power of darkness.

And now it has struck me, my darlings, that we are all, however unlike in the actual facts of our position, yet in some remarkable particulars like this Forerunner.

I need not tell you, how we are all put here to point out and testify to the same Great One who is to come—this, thank God, we know, and I believe we are doing it, each in our place, as God has given us power.

But the point in which I would make the comparison is this. We are living, as he lived, each of us, at the end of the night, at the break of a brighter day. "The night is far spent, the day is at hand." Whatever we may be now, the greatest, wisest, holiest of us all in the Church on earth is less than the least on that other side.

And we, in our imperfect estimate of our Lord's ways, may like him be disposed sometimes to look out of our earthly prison, and in displeasure at his delay, almost question his mission.

So do temptations and so do duties repeat themselves, and characters even the most unlike cover common ground.

And another thing has struck me while we have been concluding, with reference to our future subjects.

This martyrdom of John seems, as I said, to have been the first great introduction to our blessed Lord of the approaching period of his sufferings.

How will it be, if we devote some of these our evenings in tracing out his sufferings? I don't think that exactly this has been done before—I mean following out, from the very beginning, the tokens of the coming Passion. It will be work of the deepest interest, and will, I think, admirably suit these our evening talks, which are rather exercises of the fancy about divine things than regular treatments of divine things themselves.

I see Jessie is looking at the clock and at her music-stand. Well, there can be but one thing to-night—that glorious opening of the Messiah again. I wish we could do Wise's grand anthem, "Prepare ye the way." But we have not numbers for that: numbers, I mean, who can take up points such as there are in that anthem.

XI

IN these our Sunday evenings' talks we have hitherto dealt exclusively with the history of our Blessed Lord. And there is plenty yet remaining for us in that inexhaustible storehouse to draw out and enjoy hereafter.

But it may be well for us all, and even for our little one here, that we should sometimes talk over other points of religious interest. And Jessie has asked me a question, arising out of what she has been reading in the papers, which may I think profitably employ us to-night. Besides touching a point of Christian conduct on which we all want putting right, it actually does connect itself with a portion of our Lord's life on earth, and with a memorable utterance of his sacred lips.

She wonders how we are to reconcile our Lord's own sayings about his disciples with the conduct of his disciples now. And she especially selects that one saying which He uttered about the man who did not follow after the Apostles, and yet worked miracles in his name.

Well, certainly if we wanted an extreme case of nonconformity, we could not have imagined a stronger one. For here were the chosen company, and in the midst of them the Blessed One Himself, going the way appointed for Him, nay at that very time setting out on the great final journey which was to lead to his sufferings and death. And here was one, following Him not as they followed Him, having some peculiar belief and notions of his own about Him as the Messiah, yet strong enough in that faith, such as it was, to work miracles in his name. And this man goes his own way in his nonconformity and per- versity,—lets the great and holy company

round the Lord Himself, the future founders
and pillars of his Church, go their way, and
walks on in his separatist pride and irreverence.   Was there ever such a case for an
anathema ?   What a noble and decisive example might the Lord have set at that moment, which should have put down for ever
all such independence of thought and action
through the coming ages of the Church !   He
had lately rebuked an Apostle with " Get
thee behind me, Satan :" has He no words
in reserve, even stronger than those, whose
terrible sound may echo down the spaces of
time, and wither back during the long hereafter the first springings of dissidence among
his followers ?

But as we look at the case, there appears
in it something more than we have as yet
touched.   We have already remarked, that
this happened at the outset of that great last
journey to Jerusalem.   Now about that journey there was something very solemn, and of

its own kind. It was the especial dedication
of the Lamb of God to his sufferings for the
sin of the world. The resolve, uttered in the
words, " Behold, we go up to Jerusalem,"
was uppermost in his soul, when the rash
Peter rebuked Him for what seemed a morbid
dwelling on the sufferings about to be sought
by Him. It was the abhorrent idea of any
check in his own soul to that all-absorbing
and holy resolve, which led Him to turn so
sharply on his Apostle, and to designate him
as no other than a servant of the adversary
who would turn Him away from his chosen
work of love. And then in the power of that
great resolve, He went before his chosen band,
with his face stedfastly set towards Jerusalem,
divinely glorified for his divine errand. What
words are those of St. Mark, where he says,
" Jesus went before them, and they were
amazed ; and as they followed, they were
afraid " !

It seems the chapter of the Gospel history

fullest of the Divinity, fullest of the atoning Love of the Blessed Jesus: the one time of all others when those who followed behind Him were the most closely knit into his personality as the Holy One of God, most sprinkled by anticipation with the blood of that sacrifice, which He was to offer for them and for all.

And yet, at such a time, when God manifest in the flesh was bursting the bands of his humiliation, when the holy shadow of suffering was enwrapping the mystical body in union with its consecrated Head—at such a time, was one calling himself Christ's, who looked askance upon the one company which carried the world's salvation, and refused the appointed path by which the Son of God was ascending to his deed of love.

Again, what nobler opportunity could have been offered to the Lord, of once for all putting an end to the rejection of holy doctrine? That indignation with which He repelled the idea of backwardness to suffer on his own part, shall it not break forth against one who has no sympathy with, and cares not to travel on, that journey of redeeming love?

But all is not yet exhausted. Those Twelve, and the others who accompanied Him—if they imperfectly apprehended the great things of which I have been speaking, at least were ardent and blameless in their love for Himself and reverence for his person. Once, He had said something too hard for their understanding. In consequence, many of the disciples had left Him and walked no longer with Him. And when He turned to the Twelve and asked them, " Will ye also go away ?" they had replied, " Lord, to whom shall we go ? Thou hast the words of eternal life." Now it requires no stretch of imagination to persuade ourselves that this man of whom we are speaking had split off at that very time. We are not told, mind, that those who did so ceased to be disciples, but that they ceased to *walk* with Jesus. If this were so, then there

had been a distinct offence given, and an act
of alienation had taken place from the person
and company of the Divine Master. And
when we think of this, think at the same time
of all the loving invitations to come to Him
and follow after Him which the Lord was
ever giving, and then picture to yourselves
this man, who refused them all, and preferred
not to walk with Him. For such an one, can
any rebuke be too decided, any discourage-
ment too pointed? But even one thing more
remains. If—and there can be no reason why
it should not have been so—this man were one
of those who ceased to walk with the Lord on
that occasion, why and on what account had
he so gone back? I said, because Jesus had
uttered something too hard for his compre-
hension. Now, of what kind had that hard
saying been, and on what subject? Why,
precisely on the one most solemn subject on
which our Lord could discourse to his disciples—on that eating of his Flesh, and drink-

ing of his Blood, by which we have part in Him, and live by Him : that partaking of Him, which is represented to us, and made real by our faith, in the holy ordinance which He afterwards instituted—the Sacrament of the Lord's Supper. This it was, which those who went back, and walked no longer with Him, found too hard for them, and could not bear. Those it would be, who, when the ordinance pointed at by our Lord's discourse had been founded, might indeed be willing and glad to partake of it as a touching remembrance, but would still take offence at its higher and holier aspects. All the · great foundation doctrines which centre in that holy Sacrament were put aside by this disciple who walked no longer with Jesus. He saw no beauty, no life, in them : the sayings concerning them offended Him, and formed no part of his religion. Once more, and for the last time, who so richly seemed to deserve strong rebuke as this neglecter, this despiser, of the grandest doctrines which lie at the root of the Christian life ?

Let us just sum up what we have been gathering together. Here is a man calling himself a disciple of Christ. But he follows not with that band of chosen ones of which He is the centre, and who are bound together by the confession that He is the Holy One of God : he has no sympathy with that road to the Cross, along which the Lamb of God is leading his own : he has been offended at our Lord's words, and has broken away from his company ; and the words at which he has taken offence are the very character of the central doctrine of the Christian life, the partaking of the Body and Blood of the Lord in the Ordinance of his own appointing.

And now the indignant disciples, having seen him in this his separation naming the name of Christ, approach the Master with the tidings, and in look and tone demand the anathema so justly due. They themselves had

already forbidden the intruder : forbidden him
on the usual ground—" because he followeth
not with us." There could be but one way of
following Christ; and if perchance they were
disposed to allow more than one, yet it would
be impossible that this one, with all its defects
and all its offences, could be allowable.

But what is the Lord's answer ? " Forbid
him not." Oh wonderful reply ! And the
reason is not less wonderful : " For he that is
not against us is for us." To what do they
amount when united ? To what but this—
Put no check on efforts for Christ, however
ill-judged, however self-willed, however imperfect as to grasp of doctrine or entire alle-
giance to Him as others judge it : He who
knows the heart, let Him be judge of the
heart's intents. Aim not at, expect not, con-
formity : all think not alike, all feel not alike,
all act not alike. Christ came for all, is wide
enough for all, is deep enough for all : let all
come to Him—let all work for Him. Oh who

can tell the depth and breadth of love which
is in that human heart of our Blessed Master?
Very few, my darlings, make the slightest
effort to conceive it. Look at what Christ's
Church in general has done. She has laid
down a certain framework of outward govern-
ment and of doctrine, and then has faced about
to the world, and proclaimed, This or none!
All who, calling themselves by Christ's name,
doing good in Christ's name, follow not with
her, she has severely and ruthlessly forbidden.
From the first days downwards, this has been
her habit, and is her habit still. How far she
has put it forth in practice, has depended
simply on how much power the human laws
of this or that country have given her power
to persecute.   Death, torture, banishment,
spoiling of goods, deprivation of civil rights,
these have been her treatment of those respecting whom her Lord has commanded,
" Forbid him not." And where these were,
owing to good laws made by Christian states,

not in her power, there she has done all she could in the same direction. She has drawn up canons, as they are called, full of curses of such her fellow-Christians; she has called them by all sorts of hard names; she has turned the cold shoulder upon them, and excluded them from common society. And if any have made a real attempt to return to the example of our Blessed Master in their treatment of these their brethren, the persecutors have turned upon them with a rage lamentable to behold.

Let us look for a moment at the incident out of which Jessie's question arose. A body of men, associated for a high and holy work,* purpose to begin that work by receiving together the blessed Supper of the Lord. Among them are several who, living and working in Christ's name, follow not with us. To their lasting honour, they accept the proposal, and without scruple come with us to the Lord's

* The revision of the New Testament.

Table, conforming to our Service, adopting our posture, receiving the bread and wine at the hands of our minister. Was ever a scene over which our Blessed Master's loving heart would more entirely have rejoiced? Was ever a concourse which ought to have been viewed with more unmixed and ardent thankfulness? And yet, what has happened? I am not surprised, Jessie, that on reading that disgraceful page in our English Church history, you should have asked me the question which I am endeavouring to answer. Instead of thankfulness, we have had a shout of indignation and rage from the persecuting party. First of all, they were not ashamed to quote, as against a great and noble act of Christian love, the minute rules of detail which are laid down to guide us in our regular action with our own members. Then, when this would not do, they betook themselves to the assumption that some one among these our brethren denied the Divinity of our Blessed Lord. The assumption happened to be a false one : no matter : whether it were true or false, they had taken no pains to examine : it served their purpose to rouse angry feeling and to build anathemas upon. Had it been as true as it was false, we have seen what line of duty our Lord's example points out to us. From that Table Christ's disciples are not to be driven away. Do they call Him Master? Do they devote themselves to his work? Then, however little they may hold of what we ourselves esteem essential doctrine respecting Him, we have no right to exclude them. To do so, would be to insult Him, who alone knows their hearts. To do so, would be to scandalize the great yearnings of Christian humanity, which in all its thousand varieties of character, is drawn after Him. "Come unto me all ye that are weary and heavy-laden," He stands and cries. "You shall not come," shout our zealous friends. To which will the Church listen ?

Well, I suppose we have no doubt about the answer. I suppose we, though some of us are young, are old enough to see through the angry nonsense which has been talked about this matter, and to know that the Lord didn't mean his Apostles to cease from walking with Him when He told them not to forbid those who had thus ceased. We know that it is not because we value Him little, but because we value Him much, that we insist on embracing as brethren all who call themselves by his name; because we hold his blessed and adorable Person to be the centre and heart of all our faith, and not any system of truths which has been built up around that Person.

But Jessie asked me one more question, which I suppose I must try to answer also. She wanted to know how it was that our Fathers the Bishops had not answered the persecuting memorials with more earnestness and straightforwardness.

Well, Jessie, I lament as much as you can

that they have not taken the great opportunity which God has put in their way, of standing in the front of his Church and proclaiming that the same mind is in them which was also in Christ Jesus. But, my lassie, you don't know bishops as well as I do. Many of them, I believe almost all of them, feel just as you and I do on this matter. But when a good Christian man gets made a bishop, there is a great chain put on his tongue, and he can't say a great many things which he could and did say before. You must allow for this. A bishop has not merely his own view to maintain, but he has the whole Church to govern. If I had been a bishop, I never could have said what I have said to you to-night. Half my diocese would have been up in arms, thundering, foaming, memorialising, writing to all manner of newspapers.

So I suppose we must be contented as things are. For my part, I am most thankful to have got the answers we have. Our fathers

have been led to commit themselves to some
noble truths which they had not openly confessed before, and which we, who are not
bishops, shall henceforth take our stand upon
as acknowledged principles.

And that surely is no small gain.

But Margey is fast asleep on my knee, and
the servants will be wondering why the bell
doesn't ring for prayers; so just get the books,
Jessie, and look out

"All people that on earth do dwell."

XII.

THIS evening has been bespoken by our little Margey. And as she hides her head under my arm, and won't plead her own cause, I suppose I must say for her what it is which has made her wish for to-night to be hers. So, little one, I am going to tell your story for my text, and then to give you the Homily which will follow.

Margey tells me she went to sleep last night with the bright full moon shining in at her window. The light on the blind was so white, that she jumped out of bed and drew it up; and she could see the dark wood at the top of the hill over the glen, and the white sheep in the square fold, just as we watched the shepherd put them in last night, and all the grass and shrubs cut in level lines with the silvery mist. And as she lay long afterwards, the great moonbeam filled half the room, and sparkled on the water-bottle, and threw a little glory on the wall from the looking-glass. So she slept— and had a dream. And the dream was all strange and chopped up and confused, and — yes, Margey, I feel the nudge—she had rather I wouldn't tell it. But the "outcome" of it, as some of our friends say, is this: that she wants me to give her an evening about the Transfiguration.

Well, first let us remember a little of what we said last Sunday about a certain time in our Lord's life on earth : for it is near to that same period that the Transfiguration belongs. I have often thought of the Transfiguration as like a guide lighting a candle when he is about to take people into a dark place. He himself knows the course perfectly well: every step, and every stone, and every treacherous pool ; but they do not. And therefore, less for himself than for them, he takes a light, to show them the way and make it cheerful for them. Now our Lord, at this period of his ministry, was just at the entrance of a very dark place indeed. As we saw last Sunday, He had just begun to warn them that He was going up to Jerusalem to suffer and die. They, poor souls, had no mind for his words, no feeling of their meaning. They were to them so much damp thrown on their glowing hopes; but the hopes glowed so brightly that the damp was thrown off or absorbed. Still, think what they would, the dark place was before them, and into it they must enter.

And of these disciples there were three, whose lot it would be to enter deepest and farthest—to see the power of darkness in its blackest gloom, and witness the lowest depression of the Lord's human spirit.

All then, and these three more especially, might need a candle in the gloom : might be better and firmer for some recollection which should connect their Master's glory with his suffering—which might enable them by looking back on it to reassure themselves that the rejected of earth was yet the acknowledged of heaven.

And this was not all. One of these three had lately confessed Him to be the Holy Son of God, the Christ, the Son of the living God. Now this was a mighty stretch of faith in Peter, and so was it treated by our Lord. But the same Peter, on the first mention of his coming sufferings, had taken Him and rebuked Him, declaring that this should not be to Him, and had incurred thereby his severe displeasure. And though these things were said by only one, yet doubtless the Three, as leaders of the Twelve, and the Twelve generally, had both the same firm faith in his Godhead, and the same aversion from the idea of his sufferings.

Not only then to cheer and guide, but also to convince them, was some such manifestation

needed as was given to them at the Transfiguration.

And now, my little one, we have come to
the point which you have been waiting for,
and shall soon be dealing with that of which
your dream reminded you.

St. Luke's account of what took place is the
fullest. He tells us that the Lord took the
three up into the mountain to pray. He
seems, and no wonder, to have been very much
in prayer about this time. His hour was fast
coming on, and his soul was more than ever
gathering strength for what was before Him.

It has been observed that St. Luke is especially fond of telling us that Jesus *was praying*. Thus he has this particular (iii. 21)
when He had come up out of the water at his
baptism: again (v. 16), when the fame of his
healing was spreading everywhere, we read
that He withdrew himself into the wilderness,
and prayed: again (vi. 12) we read of his
going into the mountain to pray and con-

tinuing all night in prayer to God : again, just before the present incident (ix. 18), we have Him praying alone and the disciples coming to Him : and after this (ch. xi. 1), the disciples, when He had been praying, ask Him to teach them how to pray. All this is deeply interesting. It shows us that this his constant habit made a deep impression on the eye-witness, whoever he was, to whom St. Luke was indebted for the chief part of his collected narrative.

Well, then, it was no unusual thing for Him to go away with his disciples, or with some of them, for the purpose of prayer. So He took them into the mountain, says St. Luke, and tells us no more about the place. But it is in places generally that St. Luke's narrative is not particular. St. Matthew and St. Mark both tell us that this happened near Cæsarea Philippi, far away to the north of our Lord's usual haunts by the Sea of Galilee. Our friend Dean Stanley, than whom there hardly is a better guide in such matters, thinks that the glorious snowy height of Hermon was the spot. In both St. Matthew and St. Mark it is "an high mountain," and the Dean believes that to be the only one in all Palestine which deserves such a name.

Who can tell their converse as they went up steep after steep? It was evening. Perhaps the great orb of glory was sinking to his rest far in the sea-line to westward, and the ridges of snow above them were rose-bathed with his parting ray. Beneath them are fading into mist the towers and hills and forests of one of the grandest views in the world. And those four—One leading, as with a purpose, the rest following up the stony track, till at length they reach the determined spot for prayer.

Was the prayer aloud? Did the Lord go up to pray *with* his disciples? Or was it to commune all night with his heavenly Father? By what St. Luke says, this seems most probable. For he tells us that the three were heavy with

sleep, but they kept awake, and saw what
happened. So that I suppose they were lying
or sitting, as we know they were in Gethse-
mane, at some little distance from Him, and
He was rapt in prayer, possibly, but hardly
probably, uttering words which they could
hear.

They were drowsy, and perhaps, after the
manner of drowsy persons, listlessly watching
Him, one lifting up his eyes now and then,
when suddenly something unwonted calls their
attention. They could see his Face in the dim
light turned towards heaven. But now it
seems as if some new day had arisen and was
reflected from it. Not so—it is itself the day!
From every feature light beams forth. Nor
only so: his whole Form becomes a pillar of
light. The garments of earth have become the
white robes of heaven! Above, perhaps around
them, slept the snow in its purity. But suddenly, its purity has died back into dimness.
That raiment is white as the light, exceeding

white and glistening, so as no fuller on earth can white them. And that Face, day by day known to them, it is strange and unendurable, for it is as the sun—pulsing, swimming in glory. Is it earth, or have they been translated into heaven?

Well might they doubt, for now borne visibly through the night are present two celestial forms, known to them by the loftier spiritual instinct into which they are rapt;— the Servant of God who spoke with Him face to face in the cloud on Sinai,—the Prophet of Jehovah who was borne up from Jordan in the chariot of fire. These appeared in glory. Was their glory like his? Theirs, the radiance of redeemed ones, like his, a ray of the glory which He had before the world began? We cannot tell.

But they are speaking: three celestial beings holding discourse in their shining robes and with their countenances beaming with bliss. They are speaking—of what? What words

N

are fitting for such a scene of splendour? Oh it is a lesson—a lesson to all splendour—a lesson, not to teach that there shall be no splendour, but to teach splendour why it ought to shine—to teach pride what it ought to boast of. They spoke of his decease which He should accomplish at Jerusalem. The very theme for touching which one of those three had laid hands on the Son of man and rudely chidden Him with his "This shall not be to thee;"— the very words which their hopeful spirits refused to hear or understand;—these are sounding amidst the dazzling light of those wonderful forms. As they veil their eyes and gaze, they may hear strange mention of the scourge, the crown of thorns, the cross, the three hours of darkness, the centurion's spear, the counsellor's grave.

The sounds were incongruous, they impressed them little perhaps at the time; for the rash and ready Peter, ever first to speak the impulse of the moment, finds it good for them to be there; thinks it well to detain the celestial personages, well to perpetuate the transfigured state of his Lord; wants to build three booths, or tents, or shrines, for the glorified ones to dwell in; "not knowing what he said," says St. Luke; "for he wist not what to answer, for they were sore afraid," says St. Mark, writing perhaps with Peter himself at his side. That is, joy, and excitement, and fear, were mingled, and he uttered, as so many of us do, words without meaning.

But his words passed away, and the great Vision went on. A bright cloud came over the mountain top—bright, not only with the reflected glory of the Blessed Ones, but bearing a glory of its own, even the glory which came down on Sinai, the glory which appeared between the cherubims, the glory of Jehovah Himself. There was that in it which made them tremble as they entered the cloud. Even so had one of those glorified ones trembled when called into the cloud on Sinai: even so had Jehovah answered him by a voice. But now it is not "Thou shalt" and "Thou shalt not;" it is not "These are the judgments which thou shalt teach them;" no — it is another mountain and another time, though it is the same Voice: it is the fulness of time: the end of the law is come, the desire of the prophets is come: "This is my beloved Son, this is my Son whom I have chosen; hear Him."

And the cloud lifts off, the vision passes away. Not for Moses, not for Elijah, shall the disciples have to build: Jesus was left ALONE; for Him they will have to build: not on Hermon, not on Zion, but in themselves: for there is the place where He will dwell: in their hearts by faith.

But we have been missing one main object which doubtless this great event served; the mysterious comforting and strengthening of the Lord Himself for what lay before Him. If a painful time were coming for any of us, say a

great trial in a foreign land, where none under-
stood or cared for us, what a blessing would it
be to see, if but for an hour, some dear and
valued friend from home, who might talk over
our coming trouble! And we need not be
afraid of likening the Lord to ourselves in this
matter. I cannot doubt that He came down
from this mountain strengthened and refreshed
in soul for his wonderful work of love; that in
desertion and dejection that hour of glory and
that holy converse came back upon Him and
cheered his spirit.

Many years after, one of the three had occasion to write of the truth and genuineness of
the testimony which he and his colleagues had
borne to his divine Lord. It was no clever
fable, he said, which they had followed; they
had seen his majesty with their own eyes—
they had heard the voice from the Excellent
Glory bear witness to Him. And if you
turn to that place (2 Pet. i. 16) you will find
a remarkable and pleasing trace of what

passed at the Transfiguration. Then, Peter
had wanted to build lasting *tabernacles*, and
had instead heard of the decease, the *exodus*,
the pilgrimage, of the Lord : now, he tells
them, in connection with his mention of the
heavenly vision, he is looking forward to
his own *exodus*, his decease ; for that the
Lord had shown him that he must shortly
put off this his *tabernacle*. And these words
lead him on to speak of that night on Mount
Hermon.

Well, Margey, it's a grand story : and I don't
wonder at the bright moon, and the sparkles,
and the silver mist, raising its image in your
sleeping thoughts.

From your dream we have gone back to the
past : and we might have gone forward to the
future.

Oh what will it be, dear ones, when the hills
of the blessed land shall be crowned with
transfigured ones, all appearing in glory, all
walking in light ! Much will then be changed

—but not those who meet—not the subject of their converse.

There will be those who heard Moses and the prophets,—the Church in the wilderness, the Church in Shiloh, the Church in Jerusalem, the Church in Babylon : there will be the first-fruits of Pentecost, the harvest of the centuries, the gleaning of the latter ages. And in the midst, the glorified Lord : wearing, not the crown of light to be changed for the crown of thorns, but the glory which He had before the worlds and shall never put off. And the decease accomplished at Jerusalem shall still be the theme : infinite in its interest, unfathomable in its depths.

And now suppose we read that account in St. Luke, and sing to the grand old tune the hymn whose refrain is " Crown Him Lord of all."

XIII.

AND now to-night for a reverie of my own;
like your dream, Margey, a creature of
circumstances; like your dream, graven deeply
on the mind.

It was in the hot dry South—no matter
where; we needn't plague you with a geo-
graphy lesson—that mamma and I were to go
to see a strange old place up in the mountains.
All had been ordered the night before, and the
morning rose cloudless and glorious. At an
early hour we were up and had breakfasted,
and at the appointed time we were told (and
indeed discordant sounds were heard, as usual,
betraying it) that the donkeys were at the
door.

In the evening, when we went to give our

order, we had been introduced to our guides that were to be.

Jacopo, a lad of about eleven, was to accompany me, and his sister, Tina, a girl between nine and ten, was to go beside mamma. They knew every inch of the way, and Jacopo was a marvel of steadiness for his years.

This morning, though we had heard the two voices from our room, not the children, but the mother, appeared, with something hanging over her arm. She beckoned to them to come forward, saying she hoped the Signorina would excuse the children going as they were " *e troppo caldo per portare i vestimenti,*" and in the heat of summer the children, about there, always were as we now saw these. But if we wished it, she had some more clothes which they could put on.

They were dressed simply in a little skirt or kilt, and nothing else ; and as we looked at their smooth tawny forms against the deep blue sky, we thought of those charming pictures of Dobson's, which Jessie and mamma have so often admired, and from one of which a little friend of mine now clasping my hand could not turn herself away on her first visit to the Academy last May. And we told the padrona, to the evident joy of the children, that we couldn't think of making them uncomfortable, and hoped they would go as they were.

So we set out: up steep rugged rock paths, and slippery paved roads, with the two little guides trotting by us, now chattering and playing with one another, now admonishing, with noise and otherwise, our dull reluctant beasts: the scene being as Oriental as Europe could make it: flat-roofed houses, a burnt arid vegetation,—prickly pears, and aloes, and figs, and now and then a stately palm, and the splendid vine rambling over all the fences and terraces. And then we would pass through woods of olives with their gnarled trunks and silvery green foliage, and under them wheat and lucerne, and bright pink cornflowers, which little Tina was fond of running and picking, and bringing to mamma for her hat.

At last Jacopo, with all the air of head guide, came up to mamma's donkey, and asked where the Signorina would like to rest. It was about *mezza via*, and, as he said feelingly, putting his hand on the heated medal which hung on his chest, *molto caldo*. He knew of a shady spot, a little off the road, with a spring of delicious water, and there we could lunch. I needn't say we were only too glad to close with the proposal.

Now I have brought you all to this spot, because it was there that my half-waking dream was suggested. We had discussed the contents of our basket, and shared them with our cheery little friends, while the animals enjoyed their meal browsing on the rank verdure round the spring. At last they lay down, and we followed the example. And now occurred that which set me dreaming.

Jacopo and Tina had settled themselves to sleep on the soft thymy grass. He lay, with his arms stretched over his head, one leg drawn up, the other straight, while she had thrown herself with her head on his bosom, one arm under his shoulder, the other across him, and holding the medal, with which she had been playing. I first drew mamma's attention to the group, and then thought, and thought, and closed my eyes, and thought on, till the limit was passed between the thoughts which we make and the thoughts which make themselves,—and thus they went.

"Suffer them to come to Me." It was on some such track as this, while He was teaching and healing, that the parents brought the children to the Lord. Some such burning sun glared down on the group, and over soft brown limbs like those lying twined there in childish sleep, did He pass his blessed and blessing hands.

Suffer them to come to Me. And how many

ways are there to Him? One way was that of
those Judæan parents; simple enough, if we
come to think of it. They wanted Him to
touch them. Well, this perhaps was superstitious—foolish. So thought the wise disciples. The earnest Peter, the zealous James
and John, Thomas, slow of credence, Philip,
ambitious of the Divine vision, Matthew
versed in this world's money matters,—and
that one who never had a heart for his
Master's career at all,—all these combined
with the rest in keeping those fathers and
mothers away. But what thought, and what
said He? "Bring them not for the touch
that passeth away, but bring them to be
taught, when they can learn;" was this what
He said? It was not: what He said was,
"Suffer them to come to Me": to come and
be touched, if they will: to come as they will,
as they can.

And thus the thought bore itself in upon
me, how the Lord's words are larger and more

glorious than ever we suspected them to be. I had been talking to Jacopo about such things, and I found that he, poor lad, had been taught to shrink from the very thought of Him who said these words. He hoped the Madonna would intercede for him, for he had been told that Gesú would judge him at the last day, and he had seen over the great door of the church the Judge going to throw lightning among the assembled crowd, but the kind Madonna was holding back his arm. And Tina, she was near death when she was little, and she had been brought and laid in the arms of the great Madonna in the church of San Matteo, and so she was always to belong to the Madonna, and would one day go and be a little servant girl at the Convertite Convent. But not a string of either of the little hearts had ever vibrated in harmony with a word or a look of the Blessed One.

Well, well: it is a very by-lane kind of way this of coming to Him: foolish, superstitious.

So we think: and doubtless we have, as the Apostles had, some right on our side. But shall we not listen for HIS voice? Are there no flowers of his planting in the by-lane, though the children may not meet Him there in person? May not little Tina learn from her devoted lot womanly purity, sweet thoughts and deeds of charity? May not Jacopo, even now steady beyond his years, be snatched from the burning flame of temptation, and learn to walk in the ways of One whom yet he knows not as some know Him? Shall we, in our turn, rebuke those that brought them? Shall we whisper a passing word, to break the spell of the childhood's faiths? Listen—the voice seems to come over the blue sea lying in the haze of noon—the gentle words seem to whisper in the leaves of that vine under which the soft limbs are resting—" Suffer them to come to Me."

And then the thought took another turn. How many dear children, by way of being

brought to Him, are most effectually kept
from Him? Horror of horrors used to be to
me, in my childish days, that saying of the
Catechism on Sunday—that learning of the
Collect, and proving it by texts of Scripture.
And if this was so with me, surrounded by
kindly influences, what must all the fierce dis-
cipline be to thousands of children, whereby
folk try to make them religious? " Suffer
them to come to me—forbid them not!"
They have their own ways, curious and foolish
some of them : let them come. Some play
at church, with chairs for congregation and
music-stool for pulpit, and nightgown for
surplice : I remember once playing at heaven!
And then there are all the quaint questions
and queerly put original thoughts, by which
children try to approach Him who is the great
centre of all questions. To us, they savour
of irreverence : but not to them : the weaned
child may play in the hole of the asp: forbid
them not. If it thaws, God, they say, has

used up all the frost : if it thunders, the angels are beating God's big drum. So they come : little faltering steps : shrill voices, crying Hosanna after their fashion in the aisles of the Temple : tiny hands, clapping, clapping, when they should be folded in prayer. Never mind : let them come ; forbid them not.

And sometimes while they are coming, the blessed Hand itself, even now, is laid upon them and fetches them home ; the blessed Voice, as of old, whispers, " Talitha, cumi," " Come, my child." And the journey up looks but blank to us who are below : for the little rosy cheek grows blanched, and the light of the bright eyes is dimmed, and the merry voice sounds sad and far away, and there are soft steps and watching. Again, let them come, forbid them not. We see but the wheels and axle of the fiery chariot, all the splendour is upward : and while we mourn round the bed, the fair spirit is lifted to its place in the blissful ranks, and the pure in heart has entered on the sight of God. When He calls, when He fetches them, let them come to Him.

And one thing more was borne upon me before that siesta ended and we rose and went on our way. In those wonderful Beatitudes,— and in this saying also, the Lord gave his great and mighty reason. " Forbid them not," He said: " for of such is the kingdom of God." How, and why? Just now when Jacopo and Tina were finishing our luncheon, I was watching them. The only thing in the shape of a delicacy was some sort of cake or confection, which we had just tasted, and finding too sweet, had put aside. This mamma had cut for them into two equal parts. And I saw Jacopo just play with his and eat a few crumbs off its crust, and then when Tina had done hers and wasn't looking, he quietly laid it in her lap. What did the Lord mean but this, that the fresh simplicity of kindliness, the one-aimed act of lovingness, belongs to the child, and that such we ought by his grace to keep ourselves, if we want to be of his kingdom?

Of such is the kingdom of heaven. Then, beyond a doubt, in that kingdom shall ALL the little ones be found. For it is not as children of Christians, it is not as baptized, but it is as children, that of such is that kingdom. Untainted by duplicity, by impurity, by the schemes of ripened selfishness, they are they who reflect the most unbroken rays of Him who is the Light of the world, and in them the Great Redemption takes effect at once and unquestioned.

The Gospel of the Children—how pure, how bright, how simple! It is not made up of doctrines, it has no sects, it never learned a creed. "I believe," it has never descended to: it dwells as yet in the higher realm of "I love." In it the blessed Lord is not a Personage in a book, but a shining Person, ever present, ever radiant: not one who lived

ages ago, but one seen and heard day by day.
It is the only Gospel that is written nowhere
but on the heart; the only Gospel, every one
of whose disciples shall come right at the
last.   For "of such is the kingdom of God;"
and unless we turn back from our selfishness,
from our vanity, from our duplicity, and become as one of them, there is no entrance for
us there.

So I awoke from my reverie, and the merry
voices of Jacopo and Tina were ringing round
us, and in a few minutes we were again on our
way to the strange old city in the mountains.

THE END.

PRINTED BY VIRTUE AND CO., CITY ROAD, LONDON.

www.ingramcontent.com/pod-product-compliance
Lightning Source LLC
Chambersburg PA
CBHW021706210326
41599CB00013B/1548